Lecture Notes in Artificial Intelligence   5439

Edited by R. Goebel, J. Siekmann, and W. Wahlster

Subseries of Lecture Notes in Computer Science

W0193152

Haizheng Zhang   Myra Spiliopoulou
Bamshad Mobasher   C. Lee Giles
Andrew McCallum   Olfa Nasraoui
Jaideep Srivastava   John Yen (Eds.)

Advances in Web Mining and Web Usage Analysis

9th International Workshop
on Knowledge Discovery on the Web, WebKDD 2007
and 1st International Workshop
on Social Networks Analysis, SNA-KDD 2007
San Jose, CA, USA, August 12-15, 2007
Revised Papers

 Springer

Volume Editors

Haizheng Zhang
One Microsoft Way, Redmond, WA, USA - E-mail: hazhan@microsoft.com

Myra Spiliopoulou
Otto von Guericke University, Magdeburg, Germany
E-mail: myra@iti.cs.uni-magdeburg.de

Bamshad Mobasher
DePaul University, Chicago, IL, USA - E-mail: mobasher@cti.depaul.edu

C. Lee Giles
Pennsylvania State University, University Park, PA, USA - E-mail: giles@ist.psu.edu

Andrew McCallum
University of Massachusetts, Amherst, MA, USA - E-mail: mccallum@cs.umass.edu

Olfa Nasraoui
University of Louisville, Louisville, KY, USA - E-mail: olfa.nasraoui@louisville.edu

Jaideep Srivastava
University of Minnesota, Minneapolis, MN, USA - E-mail: srivasta@cs.umn.edu

John Yen
The Pennsylvania State University, Unversity Park, PA, USA
E-mail: jyen@ist.psu.edu

Library of Congress Control Number: 2009921448

CR Subject Classification (1998): I.2, H.2.8, H.3-5, K.4, C.2

LNCS Sublibrary: SL 7 – Artificial Intelligence

ISSN   0302-9743
ISBN-10   3-642-00527-6 Springer Berlin Heidelberg New York
ISBN-13   978-3-642-00527-5 Springer Berlin Heidelberg New York

Typesetting: Camera-ready by author, data conversion by Scientific Publishing Services, Chennai, India
Printed on acid-free paper   SPIN: 12624300   06/3180   5 4 3 2 1 0

Preface

This year's volume of Advances in Web Mining and Web Usage Analysis contains the postworkshop proceedings of a joint event, the 9th International Workshop on Knowledge Discovery from the Web (WEBKDD 2007) and the First SNA-KDD Workshop on Social Network Analysis (SNA-KDD 2007). The joint workshop on Web Mining and Social Network Analysis took place at the ACM SIGKDD International Conference on Knowledge Discovery and Data Mining (KDD). It attracted 23 submissions, of which 14 were accepted for presentation at the workshop. Eight of them have been extended for inclusion in this volume.

WEBKDD is one of the most traditional workshops of the ACM SIGKDD international conference, under the auspices of which it has been organized since 1999. The strong interest for knowledge discovery in the Web, fostered not least by WEBKDD itself, has led to solutions for many problems in the Web's premature era. In the meanwhile, the Web has stepped into a new era, where it is experienced as a *social medium*, fostering interaction among people, enabling and promoting the sharing of knowledge, experiences and applications, characterized by group activities, community formation, and evolution. The design of Web 2.0 reflects the social character of the Web, bringing new potential and new challenges. The 9th WEBKDD was devoted to the challenges and opportunities of mining for the social Web and promptly gave rise to the joint event with the First Workshop on Social Network Analysis (SNA-KDD).

Social network research has advanced significantly in the last few years, strongly motivated by the prevalence of online social websites and a variety of large-scale offline social network systems. These social network systems are usually characterized by complex network structures and by rich contextual information. Researchers are interested in identifying common static topological properties of these networks, as well as the dynamics pertaining to formation and evolution. Social network analysis becomes necessary in an increasing number of application domains, including community discovery, recommendation systems, and information retrieval.

The objective of the joint WEBKDD/SNA-KDD 2007 workshop was to foster the study and interchange of ideas for the analysis and understanding of the social Web as the largest example of a social network.

Social networking on the Web is a phenomenon of scientific interest per se; there is demand for flexible and robust community discovery technologies, but also for interdisciplinary research on the rules and behavioral patterns that emerge and characterize community formation and evolution. The social flair of the Web poses new challenges and brings new opportunities for the individual. Among other things, the need for information now encompasses more than the traditional plain document search, as people started getting informed in blogs, as well as contributing with content, ratings, and recommendations to the satisfaction of the information needs of others. Data miners are expected to deliver solutions for searching, personalizing, understanding, and protecting these social structures, bearing in mind their diversity and their scale.

The WEBKDD/SNA-KDD workshop invited research results on the emerging trends and industry needs associated with the traditional Web, the social Web, and other forms of social networking systems. This included data mining advances on the discovery and analysis of communities, on personalization for solitary activities (like search) and social activities (like discovery of potential friends), and on the analysis of user behavior in social structures (like blogs).

In the first paper *Spectral Clustering in Social Networks*, Miklós Kurucz, András A. Benczúr, Károly Csalogány, and László Lukács study large graphs of interconnected entities like phonecall networks and graphs of linked Web pages. They study the potential of spectral clustering for the identification of modular and homogeneous clusters and propose heuristics that alleviate shortcomings of the basis method and yield better results with respect to homogeneity and to the distribution of cluster sizes.

In the second paper *Looking for Great Ideas: Analyzing the Innovation Jam*, Wojciech Gryc, Mary Helander, Rick Lawrence, Yan Liu, Claudia Perlich, Chandan Reddy, and Saharon Rosset of IBM T.J. Watson Research Center report on methods for the analysis of the *Innovation Jam*. IBM introduced this online discussion forum in 2006, with the objective of providing a platform where new ideas were fostered and discussed among IBM employees and some external participants. The authors report on their findings about the activities and the social formations within this forum, and about their methods for analyzing the graph structure and the contributed content.

The third paper *Segmentation and Automated Social Hierarchy Detection through Email Network Analysis* by Germán Creamer, Ryan Rowe, Shlomo Hershkop, and Salvatore J. Stolfo studies the potential of data mining in corporate householding. The task is the identification of patterns of communication and the ranking of relationships among persons that communicate electronically, in particular through email. The authors have analyzed the Enron mailserver log and compared their findings with the human-crafted knowledge about the relationships of major players in that corporation.

The fourth paper *Mining Research Communities in Bibliographical Data* by Osmar R. Zaïane, Jiyang Chen, and Randy Goebel studies the implicit relationships among entities in a bibliographic database. Bibliographic data are of paramount importance for a research community, but the understanding of the underlying social structure is not straightforward. The authors have studied the DBLP database and designed the *DBConnect* tool. *DBConnect* uses random walks to identify interconnected nodes, derive relationships among the individuals/authors that correspond to these nodes, and even formulate recommendations about research cooperations among network members.

In the fifth paper *Dynamics of a Collaborative Rating System*, Kristina Lerman studies the Web as a participatory medium, in which users contribute, distribute, and evaluate information and she investigates collaborative decision taking in the news aggregator platform Digg. Decision taking refers to the selection of the front-page stories featured regularly by Digg. This selection is based on the preferences of individual users, so the author investigates how a user influences other users and how this influence changes when a user contributes new content and obtains new friends.

In the sixth paper *Applying Link-Based Classification to Label Blogs*, Smitri Bhagat, Graham Cormode, and Irina Rozenbaum study the challenge of object labeling in blogs, thereby exploiting the links used by bloggers to connect related contents. They model this task as a graph labeling problem, for which they propose generic solutions. They then apply these solutions to the issue of blog labeling, whereby they are not only considering content but also the profiles of the bloggers themselves, attempting to assess the similarity of bloggers with respect to specific properties, such as age and gender, by studying the graph structure in which they participate.

In the seventh paper *Why We Twitter: An Analysis of a Microblogging Community*, Akshay Java, Xiaodan Song, Tim Finin, and Belle Tseng study the microblogging platform Twitter to understand the motives of users who choose microblogging for communication and information sharing. They identify four categories of microblogger intention, as well as different user roles within Twitter. They stress the differences between blogging and microblogging and compare the statistics of traffic in Twitter with those of blogs and other social networks.

In the last paper *A Recommender System Based on Local Random Walks and Spectral Methods*, Zeinab Abbassi and Vahab S. Mirrokni study interlinked blogs and propose a recommendation system for blogs that exploits this link structure. They observe the blogs as nodes of a social network, design a metric of similarity among them and devise also a *personalized* rank vector that expresses the relevance among nodes in the social network. They analyze the blog network, identify connected and strongly connected components and propose two algorithms that use this structure to formulate recommendations to a user.

August 2007

Haizheng Zhang
Myra Spiliopoulou
Bamshad Mobasher
C. Lee Giles
Andrew McCallum
Olfa Nasraoui
Jaideep Srivastava
John Yen

Organization

Workshop Chairs

Haizheng Zhang	Microsoft, USA
Myra Spiliopoulou	Otto von Guericke University Magdeburg, Germany
Lee Giles	Pennsylvania State University, USA
Andrew McCallum	University of Massachusetts, Amherst, USA
Bamshad Mobasher	DePaul University, USA
Olfa Nasraoui	University of Louisville, USA
Jaideep Srivastava	University of Minnesota, USA
John Yen	Pennsylvania State University, USA

Program Committee

Lada Adamic	University of Michigan
Sarabjot S. Anand	University of Warwick
Ricardo Baeza-Yates	Yahoo Research & Univ. Pompeu Fabra-Barcelona
Arindam Banerjee	University of Minnesota
Bettina Berendt	HU Berlin
Ed Chi	Xerox PARC
Tina Eliassi-Rad	Lawrence Livermore National Laboratory
Lise Getoor	University of Maryland
Joydeep Ghosh	University of Texas
Mark K. Goldberg	Rensselaer Polytechnic Institute
Andreas Hotho	University of Kassel
David Jensen	University of Massachusetts, Amherst
Ke Ke	Central Washington University
Ravi Kumar	Yahoo Research
Mark Last	Ben-Gurion University
Victor Lesser	University of Massachusetts, Amherst
Jure Leskovec	Carnegie Mellon University
Mark Levene	Birkbeck University of London
Ee-Peng Lim	Nanyang Tech. University, Singapore
Huan Liu	Arizona State University
Sanjay Kumar Madria	University of Missouri-Rolla
Ernestina Menasalvas	University Polytecnica Madrid, Spain
Dunja Mladenic	J. Stefan Institute, Slovenia
Alex Nanopoulos	Aristotle University, Greece
Seung-Taek Park	Yahoo! Research

Srinivasan
  Parthasarathy   Ohio State University
Jian Pei   Simon Fraser University, Canada
Xiaodan Song   NEC Labs America
Chris Volinsky   AT&T Labs-Research
Stefan Wrobel   Fraunhofer IAIS
Xifeng Yan   IBM Research
Mohammed Zaki   Rensselaer Polytechnic Institute
Alice Zheng   Carnegie Mellon University

Table of Contents

Spectral Clustering in Social Networks[*]

Miklós Kurucz, András A. Benczúr, Károly Csalogány, and László Lukács

Data Mining and Web search Research Group, Informatics Laboratory
Computer and Automation Research Institute of the Hungarian Academy of Sciences
{mkurucz,benczur,cskaresz,lacko}@ilab.sztaki.hu

Abstract. We evaluate various heuristics for hierarchical spectral clustering in large telephone call and Web graphs. Spectral clustering without additional heuristics often produces very uneven cluster sizes or low quality clusters that may consist of several disconnected components, a fact that appears to be common for several data sources but, to our knowledge, no general solution provided so far. Divide-and-Merge, a recently described postfiltering procedure may be used to eliminate bad quality branches in a binary tree hierarchy. We propose an alternate solution that enables k-way cuts in each step by immediately filtering unbalanced or low quality clusters before splitting them further.

Our experiments are performed on graphs with various weight and normalization built based on call detail records and Web crawls. We measure clustering quality both by modularity as well as by the geographic and topical homogeneity of the clusters. Compared to divide-and-merge, we give more homogeneous clusters with a more desirable distribution of the cluster sizes.

Keywords: spectral clustering, telephone call graph, social network mining, Web graph.

1 Introduction

In general, clustering covers a wide class of methods to locate relevant information and organize it in an intelligible way. The purpose of clustering telephone users includes user segmentation, selection of communities with desired or undesired properties as e.g. high ADSL penetration or high recent churn rate or for viral marketing [38]: we form groups to enhance marketing communication by also relying on the spread of information within the social network. We show, even if geographic location is available, clusters have more desirable properties such as the weight of edges across different clusters are much smaller.

The main contribution of our research is the use of telephone call graphs for testing and evaluating clustering algorithms. We believe the telephone call graphs

---

[*] Support from a Yahoo Faculty Research Grant and by grant *ASTOR* NKFP 2/004/05. This work is based on an earlier work: Spectral Clustering in Telephone Call Graphs, in Proceedings of the 9th WebKDD and 1st SNA-KDD 2007 workshop on Web mining and social network analysis, Pages 82–91 (2007) (C) ACM, 2007. http://doi.acm.org/10.1145/1348549.1348559

H. Zhang et al. (Eds.): WebKDD/SNA-KDD 2007, LNCS 5439, pp. 1–20, 2009.

behave similar to other social networks such as those of bloggers and our results may be used in a more general setting. As additional experiments we included a measurement on the LiveJournal blogger network where we showed the hardness of the spectral clustering task as well as identified the well-known Russian user group [25,47] in Section 3.4. We also partitioned the UK2007-WEBSPAM host graph where our algorithm was able to identify meaningful clusters while baseline algorithms completely failed.

The telephone call graph is formed from the call detail record, a log of all calls within a time period including caller and callee id, duration, cost and time stamp. The vertex set consists of all nodes that appear at least once as caller or callee; over this set calls form directed edges from caller to callee. Edges are weighted by various aggregates of call multiplicity, duration or cost; time stamps are ignored in this work. The resulting graph obeys the power law degree distribution and contains a giant connected component of almost all nodes [1].

We compare several clustering measures on call graphs. Unlike in the examples of [28], in our graphs the "right" clustering is by no means obvious but, similar to the findings of [28], the goodness measures can be fooled. The typical examples of practically useless spectral splits have uneven sizes or disconnected clusters; in certain cases the clustering procedure simply wastes computational resources for unnecessary steps, a phenomenon reported in particular for power law graphs [32]. We believe our findings are beyond "it works well on my data" and apply to a more general class of social networks or other small-world power law graphs.

Practical evaluation of spectral clustering in graphs is investigated mainly in the area of netlist partitioning [4] with the recent exception of the findings of Lang [31,32]. He suggests semidefinite programming techniques to avoid imbalanced cuts, however the reported running times are several hours for a single cut even for 10 million edge graphs. Techniques to scale the semidefinite programming based approaches and a comparison of the performance remains future work.

The telephone call graph appears less in the publications of the data mining community compared to the social network of bloggers [30, and references therein] or the World Wide Web [19, and many others]. Few exceptions include a theoretical analysis of connected components and eigenvalues [1,15,14] and several churn prediction by machine learning methods on real data [44,5, etc.]. Closest to our results are the structural investigations of mobile telephone call graphs [35,36] and the sketch-based approximate k-means clustering of traffic among AT&T collection stations over the United States [16]; for this latter result however the underlying graph is much smaller (20,000 nodes) and their main goal is to handle the time evolution as an additional dimension. Telephone call graphs are also used by the graph visualization community: [46] reports visualization on graphs close to the size of ours with efficient algorithms to select the neighborhood subgraph to be visualized. In addition [17] gives an example of long distance telephone call fraud application by manual investigation.

While a comprehensive comparison of clustering algorithms is beyond the scope of this paper, we justify the use of a top-down hierarchical clustering by observing that telephone call graphs and social networks in general are small world power law graphs. Small world implies very fast growth of neighborhood that strongly overlap; power law implies high degree nodes that locally connect a large number of neighboring nodes. Recent bottom-up alternatives such as clique percolation [18] suffer from these phenomena: the extreme large number of small (size 5 or 6) cliques do not only pose computational challenges but also connect most of the graph into a single cluster; the number of larger sized cliques however quickly decays and by using them we leave most of the nodes isolated or in very small clusters. The superiority of spectral clustering over density based methods is also suggested in [11] for document collections.

The applicability of spectral methods to graph partitioning is observed in the early 70's [23,21]. The methods are then rediscovered for netlist partitioning, an area related to circuit design, in the early 90's [10,2,4,3]. Since the "Spectral Clustering Golden Age" [20,34,48, etc] 2001 we only list a random selection of results. Spectral clustering is applied for documents [48,11,12] as well as image processing [41,33,34], see many earlier references in [28]. More recently, several approximate SVD algorithms appeared [24,22,39, and many others]; with the expansion of available data volumes their use in practice is likely in the near future.

Our experiments are performed on the call graph of more than two millions of Hungarian landline telephone users [7], a unique data of long time range with sufficiently rich sociodemographic information on the users. We differ from prior work on spectral clustering in two aspects:

 - We evaluate clustering algorithms by measuring external sociodemographic parameters such as geographic location in addition to graph properties such as cluster ratio.
 - Our problems are larger than previously reported: the recent Divide-and-Merge algorithm [12] runs experiments over 18,000 nodes and 1.2 millions of nonzeroes compared to near 50,000,000 edges in our graph. Improved hardware capabilities hence require new algorithms and lead to new empirical findings in the paper.

We summarize our key findings on implementing spectral clustering in the telephone call graph that may be applied for other graphs as well.

 - We give a k-way hierarchical clustering algorithm variant that outperforms the recently described Divide-and-Merge algorithm of Cheng et al. [12] both for speed and accuracy.
 - Compared to the Laplacian $D - A$ typically used for graph partitioning, we show superior performance of the normalized Laplacian $D^{-1/2}AD^{-1/2}$ introduced for spectral bisection in [41] and [20] as the relaxation of the so-called normalized cut and min-max cut problems, respectively. We are aware of no earlier systematic experimental comparison. While in [41,19] both described, their performance is not compared in practice; Weiss [45] reports "unless the matrix is normalized [...] it is nearly impossible to extract segmentation information" but no performance measures are given; finally [43] give theoretic evidence for the superiority of normalization.

- We compare various edge weighting schemes, in particular introduce a neighborhood Jaccard similarity weight. This weight outperforms the best logarithmic weighting in certain cases, justifying the discussion of [28, Section 2] that the input matrix should reflect the similarity of the nodes instead of their distance.
- We introduce size balancing heuristics that improve both the geographic homogeneity and the size distribution of the clusters formed by the algorithm. These methods outperform and completely replace Lin-Kernighan type heuristics proposed by [20].
- We partially justify previous suggestions to use several eigenvectors [4,33]; however we observe no need for too many of them.

2   Spectral Clustering Algorithms for the Call Graph

Spectral clustering refers to a set of a heuristic algorithms, all based on the overall idea of computing the first few singular vectors and then clustering in a low (in certain cases simply one) dimensional subspace. Variants dating back to the 1970's described in [4] fall into two main branches. The first branch is initiated by the seminal work of Fiedler [23] who separates data points into the positive and negative parts along the principal axes of the projection. His original idea uses the second singular vector, the so-called Fiedler vector; later variants [6,2] use more vectors. Hagen and Kahng [27] is perhaps the first to use the second smallest eigenvalue for graph partitioning of difficult real world graphs.

The second branch of hierarchical spectral clustering algorithm divides the graph into more than two parts in one step. While the idea of viewing nodes as d-dimensional vectors after projecting the adjacency matrix into the space of the top k singular vectors is described already by Chan et al. [10], much later Zha et al. [48] introduce the use of k-means over the projection.

The formation of the input matrix to SVD computation from the detailed call list strongly affects the outcome of clustering. In addition to various ways of using cost and duration including a neighborhood Jaccard similarity weight, in Section 2.5 we also compare the use of the Laplacian and weighted Laplacian. The *Laplacian* is $D - A$ such that D is the diagonal matrix where the i-th entry is the total edge weight at node i. The *weighted Laplacian* $D^{-1/2}AD^{-1/2}$ is first used for spectral bisection in [41,20]. The Laplacian arises as the relaxation of the minimum ratio cut [27]; weighted Laplacian appears in the relaxation of normalized cut [41] and min-max cut [20].

While implementation issues of SVD computation are beyond the scope of the paper, we compare the performance of the Lanczos and block Lanczos code of svdpack [8] and our implementation of a power iteration algorithm. Hagen et al. [27] suggest fast Lanczos-type methods as robust basis for computing heuristic ratio cuts; others [28,12] use power iteration. Since the SVD algorithm itself has no effect on the surrounding clustering procedure, we only compare performances later in Section 3.6.

2.1   Overview of the Algorithm

In the bulk of this section we describe our two main algorithms, one belonging to each branch of hierarchical spectral clustering. In both cases good cluster qualities are obtained by heuristics for rejecting uneven splits and small clusters described in general in Section 2.2. The first algorithm in Section 2.3 is based on k-way hierarchical clustering as described among others by Alpert et al. [4]; the second one in Section 2.4 on the more recent Divide-and-Merge algorithm [12].

When clustering the telephone call graph, the main practical problem arises when the graph or a remaining component contains a densely connected large subset. In this case spectral clustering often collects tentacles loosely connected to the center [13] into one cluster and keeps the dense component in one [32]. While even the optimum cluster ratio cut might have this structure, the disconnected cluster consists of small graph pieces that each belong strongly to certain different areas within the dense component. In addition a disconnected graph has multiple top eigenvalue, meaning that we must compute eigenvectors separate for each connected component. However if we treat each connected component as a separate cluster, we obtain an undesired very uneven distribution of cluster sizes.

Both of the algorithms we describe target at balancing the output clusters. The original Divide-and-Merge algorithm of [12] achieves this simply by producing more clusters than requested and merging them in a second phase. We observed this algorithm itself is insufficient for clustering power law graphs since for our data it chops off small pieces in one divide step. In a recursive use for hierarchical clustering the number of SVD calls hence becomes quadratic in the input size even if only a relative small number of clusters is requested.

The key in using spectral clustering for power law graphs is our small cluster redistribution heuristics described in the next subsection. After computing a 2-way or k-way split we test the resulting partition for small clusters. First we try to redistribute nodes to make each component connected. This procedure may reduce the number of clusters; when we are left with a single cluster, the output is rejected. The main difference in our two algorithms is the way rejected cuts are handled as described in Sections 2.2 and 2.3.

2.2   Small Cluster Redistribution Heuristics

We give a subroutine to reject very uneven splits that is used in both our Divide-and-Merge implementation (Section 2.2) and in k-way clustering (Section 2.3). Given a split of a cluster (that may be the entire graph) into at least two clusters $C_1 \cup \ldots \cup C_k$, we first form the connected components of each C_i and select the largest C_i'. We consider vertices in $C_i - C_i'$ outliers. In addition we impose a relative threshold `limit` and consider the entire C_i outlier if C_i' is below limit.

Next we redistribute outliers and check if the resulting clustering is sensible. In one step we schedule a single vertex v to component C_j with $d(v, C_j)$ maximum where $d(A, B)$ denotes the number of edges with one end in A and another in B. Scheduled vertices are moved into their clusters at the end so that the output

---

Algorithm 1. `redistribute`$(C_1, \ldots, C_k)$: Small cluster redistribution

---

for all C_i **do**
  $C'_i \leftarrow$ largest connected component of C_i
  if $|C'_i| < \texttt{limit} \cdot |C_1 \cup \ldots \cup C_k|$ **then**
    $C'_i \leftarrow \emptyset$
Outlier $= (C_1 - C'_1) \cup \ldots \cup (C_k - C'_k)$
for all $v \in$ Outlier **do**
  $p(v) \leftarrow j$ with largest total edge weight $d(v, C'_j)$
for all $v \in$ Outlier **do**
  Move v to new cluster $C_{p(v)}$
return all nonempty C_i

---

is independent of the order vertices v are processed. By this procedure we may be left with less than k components; we will have to reject clustering if we are left with the entire input as a single cluster. In this case we either try splitting it with modified parameters or completely give up forming subclusters.

2.3 K-Way Hierarchical Clustering

In our benchmark implementation we give k, the number of subclusters formed in each step, d, the dimension of the SVD projection and `cnum`, the required number of clusters as input. Algorithm 2 then always attempts to split the largest available cluster into $k' \leq k$ pieces by k-means after a projection onto d dimensions. Note that k-means may produce less than the prescribed number of clusters k; this scenario typically implies the hardness of clustering the graph. If, after calling small cluster redistribution (Algorithm 1), we are left with a single cluster, we discard C_0 and do not attempt to split it further.

In our real life application we start out with low values of d and increase it for another try with C_0 whenever splitting a cluster C_0 fails. We may in this case also decrease the balance constraint.

Notice the row normalization step $v_i \leftarrow v'_i / ||v'_i||$; this step improves clustering qualities for our problem. We also implemented column normalization, its effect is however negligible.

Algorithm 2. k-way hierarchical clustering

---

while we have less than `cnum` clusters **do**
  $A \leftarrow$ adjacency matrix of largest cluster C_0
  Project $D^{-1/2} A D^{-1/2}$ into first d eigenvectors
  For each node i form vector $v'_i \in R^d$ of the projection
  $v_i \leftarrow v'_i / ||v'_i||$
  $(C_1, \ldots, C_k) \leftarrow$ output of k-means$(v_1, \ldots, v_{|C_0|})$
  Call `redistribute`$(C_1, \ldots, C_k)$
  Discard C_0 if C_0 remains a single cluster

2.4   Divide-and-Merge Baseline

The Divide-and-Merge algorithm of Cheng et al. [12] is a two phase algorithm. In the first phase we recursively bisect the graph: we perform a linear scan in the second eigenvector of the Laplacian sorted by value to find the optimal bisection. The algorithm produces cnum$_0$ clusters that are in the second phase merged to a required smaller number cnum of clusters by optimizing cut measures via dynamic programming.

In order to adapt the Divide-and-Merge algorithm originally designed for document clustering [12], we modify both phases. First we describe a cluster balancing heuristic based on Algorithm 1 for the divide phase. Then for the merge phase we give an algorithm that produces low cluster ratio cuts, a measure defined below in this section. In [12] the merge phase of the divide-and-merge algorithm is not implemented for cluster ratio. Since this measure is not monotonic over subclusters, we give a new heuristic dynamic programming procedure below.

We observed tiny clusters appear very frequent in the Divide phase (Algorithm 3) as described in Section 2.2. Splits along the second eigenvector are apparently prone to find a disconnected small side consisting of outliers. In this case the small component heuristics of Algorithm 1 are insufficient themselves since we are starting out with two clusters; if we completely redistribute one, then we are left with the component unsplit. We hence introduce an additional balancing step with the intent to find connected balanced splits along the second eigenvector. We could restrict linear scan to an $1/3$-$2/3$ split; in many cases this however leads to a low quality cut. Hence first we weaken the restriction to find an $1/\texttt{ratio\_init}$-$(1 - 1/\texttt{ratio\_init})$ cut and gradually decrease the

---

Algorithm 3. Divide and Merge: Divide Phase

---

while we have less than cnum$_0$ clusters **do**
   $A \leftarrow$ adjacency matrix of largest cluster C_0
   Compute the second largest eigenvector v' of $D^{-1/2}AD^{-1/2}$
   Let $v = D^{-1/2}v'$ and sort v
   $i \leftarrow \texttt{ratio\_init}$
   while C_0 is not discarded **do**
       Find $1/i \leq t \leq 1 - 1/i$ such that the cut

$$(S, T) = (\{1, \ldots, t \cdot n\}, \{t \cdot n + 1, \ldots, n\})$$

   minimizes the cluster ratio
   $(C_1, \ldots, C_\ell) \leftarrow \texttt{redistribute}(S, T)$
   if $\ell > 1$ **then**
       Discard C_0 and add clusters $C_1, \ldots, C_\ell$
   else
       if $i = 3$ **then**
           Discard cluster C_0
       else
           $i \leftarrow i - 1$

Algorithm 4. Merge Phase

for all clusters C_0 from leaves up to the root **do**
    if C_0 is leaf **then**
        $\mathrm{OPTn}(C_0, 1) = 0$, $\mathrm{OPT_d}(C_0, 1) = |C_0|$
    else
        Let $C_1, \ldots, C_\ell$ be the children of C_0
        for i between 1 and total below C_0 **do**
            $\mathrm{numer}(i_1, \ldots, i_\ell) \leftarrow 0$; $\mathrm{denom}(i_1, \ldots, i_\ell) \leftarrow 1$
            for all $i_1 + \ldots + i_\ell = i$ **do**
                $\mathrm{numer}(i_1, \ldots, i_\ell) \leftarrow \sum_{j \neq j'} d(C_j, C_{j'}) + \sum_{j=1\ldots\ell} \mathrm{OPTn}(C_j, i_j)$
                $\mathrm{denom}(i_1, \ldots, i_\ell) \leftarrow \sum_{j \neq j'} |C_j| \cdot |C_{j'}| + \sum_{j=1\ldots\ell} \mathrm{OPT_d}(C_j, i_j)$
                if $\dfrac{\mathrm{OPT}_n(C_0, i)}{\mathrm{OPT}_d(C_0, i)} > \dfrac{\mathrm{numer}(i_1, \ldots, i_\ell)}{\mathrm{denom}(i_1, \ldots, i_\ell)}$ **then**

                $\mathrm{OPT}_n(C_0, i) = \mathrm{numer}(i_1, \ldots, i_\ell)$; $\mathrm{OPT}_d(C_0, i) = \mathrm{denum}(i_1, \ldots, i_\ell)$

---

denominator down to 3. We stop with the first cut not rejected by Algorithm 1. If no such exists, we keep the cluster in one and proceed with the remaining largest one.

Now we turn to the the Merge phase (Algorithm 4). Our goal is to optimize the final output for cluster ratio defined below. Let there be N users with N_k of them in cluster k for $k = 1, \ldots, m$. The *cluster ratio* is the number of calls between different clusters divided by $\sum_{i \neq j} N_i \cdot N_j$. The *weighted cluster ratio* is obtained by dividing the total weight of edges between different clusters by $\sum_{i \neq j} w_{ij} N_i \cdot N_j$ where w_{ij} is the total weight of edges between cluster i and j.

In order to compute the optimal merging upwards from leaves by dynamic programming (Algorithm 4) we aim to use an idea similar to computing cluster ratio when linearly scanning in the Divide step as described in [11]. Unfortunately however cluster ratio is not monotonic in the cluster ratio within a subcomponent; instead we have to add the numerator and denominator expressions separately within the subcomponents. We can only give a heuristic solution below to solve this problem.

In order to find a good cluster ratio split into i subsets of a given cluster C_0, we try all possible $i_1 + \ldots + i_\ell = i$ split sizes within subclusters $C_1, \ldots, C_\ell$. By the dynamic programming principle we assume good splits into i_j pieces are known for each subcluster C_j; as we will see, these may not be optimal though. For these splits we require the numerator and denominator values $\mathrm{OPT}_n(C_j, i_j)$ and $\mathrm{OPT}_d(C_j, i_j)$. If we use the corresponding splits for all j, we obtain a split of cluster ratio

$$\frac{\sum_{j \neq j'} d(C_j, C_{j'}) + \sum_{j=1\ldots\ell} \mathrm{OPTn}(C_j, i_j)}{\sum_{j \neq j'} |C_j| \cdot |C_{j'}| + \sum_{j=1\ldots\ell} \mathrm{OPT_d}(C_j, i_j)}$$

for the union of the subcomponents. Note however that this expression is not monotonic in the cluster ratio of subcomponent j, $\mathrm{OPT}_n(C_j, i_j)/\mathrm{OPT}_d(C_j, i_j)$, and the minimization of the above expression cannot be done by dynamic programming. As a heuristic solution, in Algorithm 4 we always use the optimal splits from children. Even in this setting the algorithm is inefficient for branching factor more than two; while in theory Merge could be used after k-way partitioning as well, the running time is exponential in k since we have to try all (or at least most) splits of i into $i_1 + \ldots + i_\ell$.

2.5 Weighting Schemes

Spectral clustering algorithm may take any input matrix A and partition the rows based on the geometry of their projection into the subspace of the top k singular vectors [28]. Kannan et al. [28] suggest modeling the input as a similarity graph rather than as a distance graph, raising the question of interpreting the call information including the number, total duration and price between a pair of callers.

Earlier results for graph partitioning either use the unweighted or weighted Laplacian $D - A$ vs. $D^{-1/2}AD^{-1/2}$, the first appearing in the relaxation of the ratio cut [27], the second the normalized [41] and min-max [20] cut problems. Weighting strategies in more detail are discussed for netlist partitioning are only [4, and references therein]; in particular Alpert and Kahng [2] empirically compared some of them. Since netlists are hypergraphs, we may not directly use their findings, however they indicate the importance of comparing different strategies to weight the graph.

We have several choices to extract the social network based on telephone calls between users: we may or may not ignore the direction of the edges and weight edges by number of calls, duration or price, the latter emphasizing long range contacts.

First of all we my try to directly use the total cost or duration as weight in the adjacency matrix. However then the Lanczos algorithm converges extremely slow; while it converges within a maximum of 120 iterations in all other cases, 900 iterations did not suffice for a single singular vector computation with raw values. We hence use $1 + \log w_{ij}$ where w_{ij} is either the total cost or duration between a pair of users i and j.

We also investigate a Jaccard similarity based weight of user pairs that characterize the strength of their connection well, based on the remark of [28] for modeling the input as a similarity graph. Since filling a quadratic size matrix is infeasible, we calculate the ratio of their total call duration made to common neighbors and of their total duration for all existing edges. This method results in weights between 0 and 1; the reweighted graph yields clusters of quality similar to the logarithm of call cost or duration. In our algorithms we use $1 + \mathrm{Jac}_{ij}$ to distinguish non-edges from low weight edges.

We build a graph from the detailed call record so that a vertex is assigned to each customer and an edge is drawn between two vertices if they have called each other during the specified time period. The edges are weighted by the total time of calls between the two vertices. However, this weighting requires quadratic space and is hence infeasible for the scale of our problem; we only compute the weight for existing edges.

This similarity coefficient is also useful for finding important connections and ignoring "accidental" unimportant connections. We sort all pairs of vertices descending by the above similarity coefficient and compare the resulting order with the actual edges of the original graph by counting the ratio of actual edges and toplist size in different sized toplists of the order. The resulting scheme downweights unimportant edges and adds "missing" calls to the network.

3   Experiments

In this section we describe our experiments performed mainly on the Hungarian Telecom call detail record and, in order to extend the scope of applicability, on the UK2007-WEBSPAM crawl and the LiveJournal blogger friends network.

The experiments were carried out on a cluster of 64-bit 3GHz P-D processors with 4GB RAM each. Depending on algorithms and parameter settings, the running time for the construction of 3000 clusters is in the order of magnitude of several hours or a day for the Hungarian Telecom data, the largest of the graphs used in our experiments.

3.1   Evaluation Measures

Graph based properties. In the next two subsections we define the quality measures we use for evaluating the output of a clustering algorithm besides cluster ratio defined in Section 2.4. While several measures other than (weighted) cluster ratio exist for measuring the quality of a graph partition, cluster ratio reflect best the balance constraints and applies best to large number of parts. We remark that we always compute cluster ratio with the original edge weights regardless of the matrix used for SVD computation.

In addition we tested *normalized network modularity* [42]

$$Q_{norm} = \sum_{\text{clusters } s} \frac{N}{N_k} \left[\left(\frac{d(C_s, \overline{C_s})}{2M} \right)^2 - \frac{d(C_s, C_s)}{M} \right]$$

where M is the total weight of the edges and $d(X, Y)$ is the weight of edges with tail in X and head in Y; we also use the notation of Section 2.4 for the sizes of the clusters. As we will see in Section 3.8, normalized modularity turned out instable and we suspect it may not be the appropriate measure for cluster quality.

Sociodemographic properties. Telephone users as nodes have rich attributes beyond graph theory. We may measure clustering quality by the entropy and purity of geographic location or other external property within the cluster. By using the notation of Section 2.4 let $N_{i,k}$ denote the cluster confusion matrix, the number of elements in cluster k from settlement i and let $p_{i,k} = N_{i,k}/N_k$

denote the ratio within the cluster. Then the *entropy* E and *purity* P [29] (the latter also called *accuracy* in [11]) are defined as

$$E = (-1/\log m) \sum_k (N_k/N) \sum_i p_{i,k} \log p_{i,k} \quad \text{and}$$

$$P = \frac{1}{N} \sum_k \max_i N_{i,k},$$

where the former is the average entropy of the distribution of settlements within the cluster while the latter measures the ratio of the "best fit" within each cluster.

3.2   Detailed Call Record Data

For a time range of 8 months, after aggregating calls between the same pairs of callers we obtained a graph with $n = 2, 100, 000$ nodes and $m = 48, 400, 000$ directed edges that include 10,800,000 bidirectional pairs.

Settlement sizes (Fig. 1, left) follow a distribution very close to lognormal with the exception of a very heavy tail of Hungary's capital Budapest of near 600,000 users. In a rare number of cases the data consists of subpart names of settlements resulting in a relatively large number of settlements with one or two telephone numbers; since the total number of such nodes is negligible in the graph, we omit cleaning the data in this respect.

We discard approximately 30,000 users (1.5%) that become isolated from the giant component; except for those 130 users initially in small components all nodes can be added to the cluster with most edges in common but we ignore them for simplicity.

The graph has strong topdown regional structure with large cities appearing as single clusters. These small world power law graphs are centered around very large degree nodes and very hard to split. In most parameter settings we are left with a large cluster of size near that of the Budapest telephone users. For this reason we re-run some experiments with Budapest users removed from the graph.

One may argue whether clustering reveals additional information compared to the settlements themselves as "ground truth" clusters. We give positive answer to this question by showing that the distribution of the total call duration across different clusters is superior for those obtained by spectral clustering. In Fig. 1, right, we form two graphs, one with a node for each settlement and another with a node for each (spectral) cluster. The weight of an edge between two such nodes is the total call duration between the corresponding clusters. We observe both graphs have power law distribution. The graph obtained by spectral clustering has a much smaller exponent and the edges across clusters have much smaller weight. In fact we use settlement information as an external validation tool for our experiments and not as ground truth.

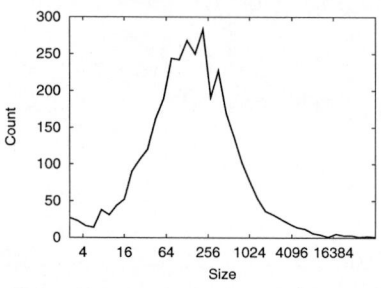
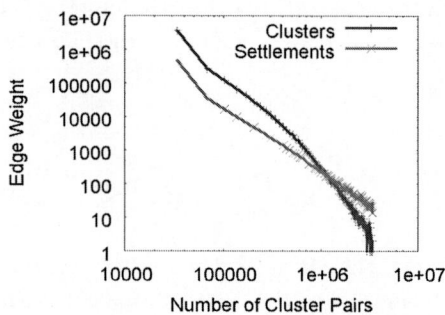

Fig. 1. Left: Distribution of the number of telephone lines by settlements in the data. **Right:** Distribution of the edge weights across different clusters, for a spectral clustering and a trivial clustering obtained by considering one settlement as one cluster. The horizontal axis contains the total edge weight in seconds and the vertical axis shows the number of cluster pairs with the given weight between them.

3.3   The UK2007-WEBSPAM Host Graph

We constructed the host graph of the UK2007-WEBSPAM crawl of Boldi et al. [9] that contains 111,149 hosts and 1,836,441 directed weighted edges. The hosts are labeled with the top level Open Directory [37] categories as in [26].

Over this host graph the k-way partitioning algorithm with $k = 15$ and $d = 30$ produced 100 clusters with three giant clusters remaining unsplit as seen in Fig. 2. In contrast, the Divide-and-Merge algorithm was not able to construct a single split, even with `ratio-init` = 8. The distribution of the 14 categories are shown in the pie charts. The first cluster has a very low fraction of known labels, most of which belongs to business (BU), computers (CO) and sports (SP), likely a highly spammed cluster. The second cluster has high ODP reference rate in business (BU), shopping (SH), computers (CO), arts (AR) and recreation (RC). Finally the largest cluster an opposite topical orinetation with high fraction of health (HE), reference (RE), science (SC) and society (SO). Among the less

hosts	labeled
23595	2622
13111	8754
38279	14964
111149	35814

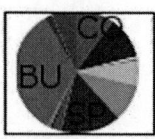

AR	arts
BU	business
CO	compouters
HE	health
RC	recreation
RE	reference
SC	science
SH	shopping
SO	society
SP	sports

Fig. 2. The size of the three largest remaining clusters and the number of labeled hosts within the cluster and in the entire crawl (bottom) as well as the distribution of categories within these clusters in the same order, left to right, with the list of abbreviations (left)

frequent four more categories, this latter cluster has a high fraction of kids and home while the second cluster contains games; news is negligible in the three clusters.

We experimented with various edge weighting schemes for the host graph, all of which resulting in roughly the same largest clusters that could not be further split as in Fig. 2, left. With an initial purity and entropy of 0.18 and 3.4, 100 clusters using the logarithm of edge weights resulted in purity 0.29, entropy 0.45, cluster ratio 0.00024 and normalized modularity -20.9, improved to 0.31, 0.44, 0.00026 and -17.9 by Jaccard weighting.

3.4   The LiveJournal Blogger Network

We crawled $n = 2,412,854$ LiveJournal bloggers and formed a graph with $m = 40,313,378$ edges, out of which 18,727,775 is bidirectional. By using $d = 30$ we obtained two clusters, one consisting of the well-known Russian user group [25,47] of 85,759 users with location given in Russia, 9,407 in Ukraine and 29,697 outside (with large number from US, Israel, Belarus and Estonia); the large cluster contains 1,849 users who gave Russia as location. When trying to split the large cluster further, we obtained an eigenvalue sequence very close to one with $\sigma_{30} = 0.986781$. For even a 100-dimensional embedding the k-way clustering algorithm managed only to chop off tiny clusters, as observed in general in [32]. Improving the performance of our algorithm for this type of data remains future work.

3.5   Divide-and-Merge vs. k-Way Hierarchical Algorithm with Different Input Matrices

The comparison of various input matrices to both divide-and-merge and k-way hierarchical clustering is shown in Fig. 3. Most importantly we notice

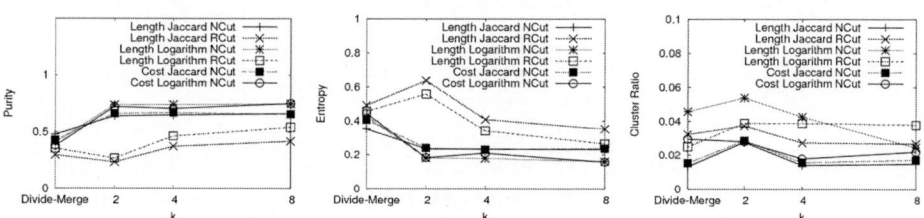

Fig. 3. Evaluation of various reweighting techniques over the adjacency matrix for purity (left), entropy (center) and cluster ratio (right) of the arising clusters on the vertical axis. Curves correspond to combinations of unweighted vs. weighted Laplacian (NCut for normalized cut relaxation, as opposed to RCut, ratio cut relaxation), length vs. cost based, and Jaccard vs. logarithmic weight input matrices. Four different algorithms, Divide-and-Merge bipartition as well as k-way partition with $d = 30$ for three values $k = 2$, 4 and 8 are on the horizontal axis.

the weighted Laplacian $D^{-1/2}AD^{-1/2}$ significantly outperforms the unweighted $D - A$ in all respects. Call length and call cost behaves similar; as expected, the former yields geographically more homogeneous clusters by underemphasizing long distance calls, while the latter performing better for the cluster ratio measure. The logarithm of the price or duration performs very close to Jaccard reweighting with no clear winner.

When comparing Divide-and-Merge and k-way partitioning (Fig. 3) we observe the superiority of the latter for larger k. For $k = 2$ we basically perform Divide without Merge; the poor performance is hence no surprise. For $k = 4$ however the small cluster redistribution heuristic already reaches and even outperforms the flexibility of the Merge phase in rearranging bad splits.

3.6   Evaluation of Singular Value Decomposition Algorithms

In our implementation we used the Lanczos code of `svdpack` [8] and compared it with block Lanczos and a power iteration developed from scratch. While block Lanczos runs much slower, it produces the exact same output as Lanczos; in contrast power iteration used by several results [28,12] is slightly faster for computing the Fiedler vector but much less accurate; computing more than two dimensions turned out useless due to the numerical instability of the orthogonal projection step. Improving numerical stability is beyond the scope of this paper and we aware of no standard power iteration implementation. Running times for the first split are shown in Table 1; in comparison the semidefinite programming bisection code of Lang [32] ran 120 minutes for a much smaller subgraph ($n = 65,000$, $m = 1,360,000$) while for the entire graph it did not terminate in two days.

Table 1. Running times for the d dimensional SVD computation by various algorithms, in minutes

Algorithm	$d = 2$	$d = 5$	$d = 10$	$d = 15$	$d = 20$	$d = 25$
Lanczos	17	24	44	47	55	96
Block Lanczos	19	34	66	105	146	195
Power	15	40	95	144	191	240

We remark that modifications of `svdpack` are necessary to handle the size of our input. After removing the obsolete condition on the maximum size of an input matrix, we abstracted data access within the implementation to computing the product of a vector with either the input matrix or its transpose.

The entire running time for producing 3000 clusters depend more on the parameter settings than the choice of Divide-and-Merge vs. k-way partitioning. All runs took several hours up to a day; only the slowest Divide-and-Merge with `limit = 100` runs over a day. Since the number of possible parameters is very large, we omitted running time graphs.

3.7   Size Limits and Implications on the Size Distribution of Clusters

In Fig. 4 we see the effect of changing `limit` for the k-way and Divide-and-Merge algorithms. Recall that in Algorithm 1 used as subroutine in both cases, the parameter `limit` bounds the ratio of the smallest cut from the average. If this is very large (100 in the Figure), we are left with no constraint. If however it is close to one, we enforce very strict balancing that deteriorates clustering quality. The optimal values lie around 4...6; these values are also optimal for running time. Very large values, in particular for Divide-and-Merge, slow algorithms down by only marginally reducing the largest cluster size after the costly SVD computation.

We checked that the strict balancing constraints required for efficiency has no negative effect on cluster quality. This is clearly seen for purity and entropy in Fig. 4, top. Notice however the unexpected increase of cluster ratio (middle left) for large values; this is due to the fact that the densely connected near 600,000 Budapest users could only be split with liberal balancing conditions as also seen in the table of largest remaining cluster sizes in Fig. 4, middle right. While splitting Budapest has no effect on purity or entropy, it adds a large number of edges cut in cluster ratio. For this reason we repeated the experiment by removing Budapest users to see no negative effect of the strict balance constraint on the clustering quality measures anymore. We did not include normalized modularity in the figures since, as we will see in Section 3.8, normalized modularity turned out instable.

Notice the superiority of the k-way algorithm over Divide-and-Merge is also clear for their best parameter settings of Fig. 4, top.

We also remark here that we implemented a Lin-Kernighan type point redistribution at cut borders proposed by [20] but it had negligible effect on the quality.

Besides clustering quality, we also look at how "natural" are the cluster sizes produced by the algorithms in Fig. 4, bottom. We observe strong maximum cluster size thresholds for Divide-and-Merge: that algorithm forces splitting hard regions for the price of producing a negatively skewed distribution of a large number of small clusters that are of little practical use. With the exception of Divide-and-Merge with no limits we never split Budapest users as seen from the top list (Fig. 4, middle right). When repeating the experiment by discarding Budapest users, the huge clusters disappear.

3.8   The Effect of More Dimensions

As suggested by [4,33] more eigenvalues produce better quality cuts. However the price for using more eigenvalues is slowdown and storage increase for the space of the vectors. Speed is shown in Table 1 while a simple calculation yields 3.2GB space requirement for 100-100 left and right vectors, stored as doubles, for 2M nodes. Hence a good balance between the number d of eigenvalues and the branching k must be chosen. In Fig. 5 we observe we should not choose k

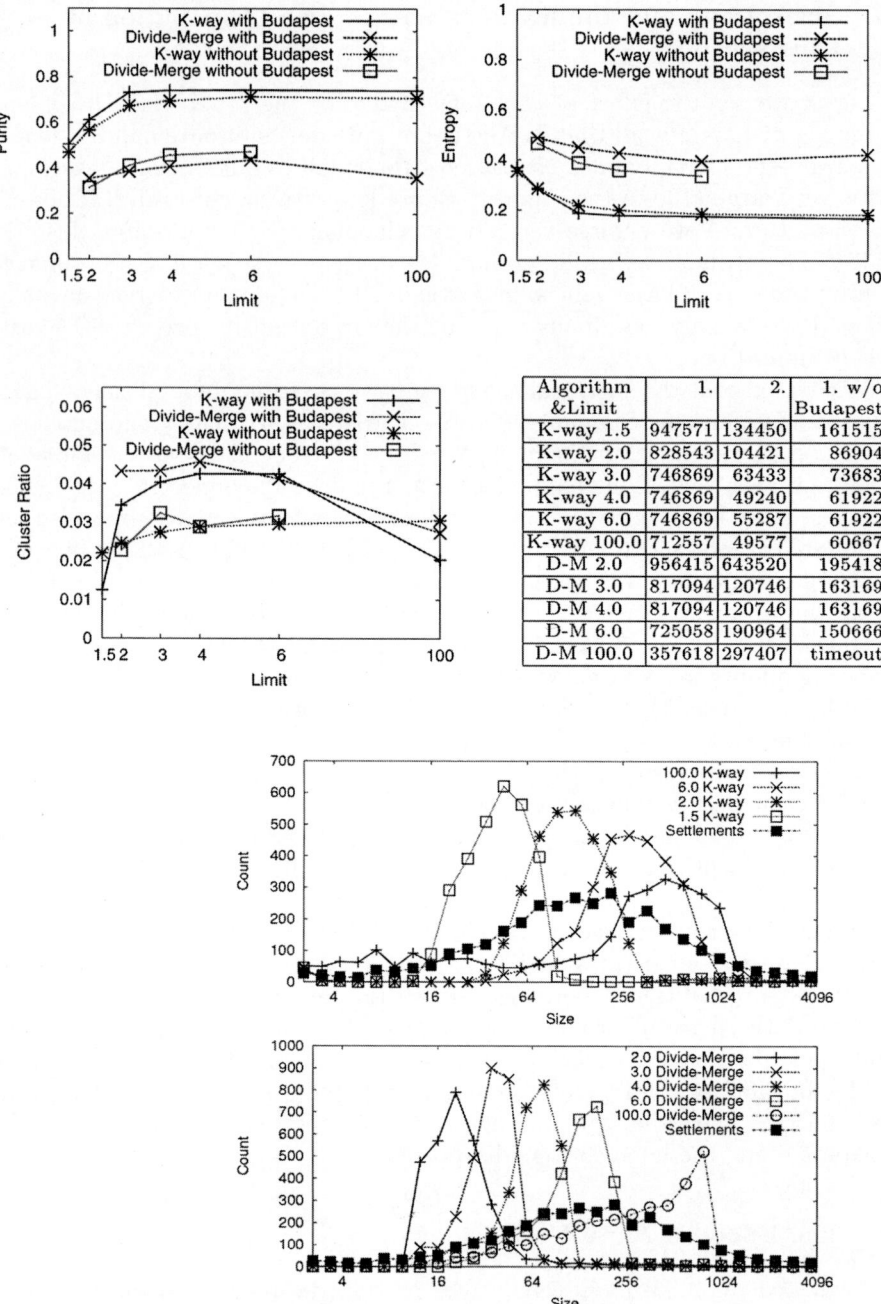

Fig. 4. Effect of size limits on clustering quality, $k = 4$ and $d = 30$ for purity, entropy (top) and cluster ratio (middle left). The size of the largest and second largest remaining cluster as well as the largest one after the removal of Budapest lines (middle right). Distribution of cluster sizes for k-way hierarchical partitioning (k-way) and Divide-and-Merge (Divide-Merge) for various parameters of the size limit (bottom).

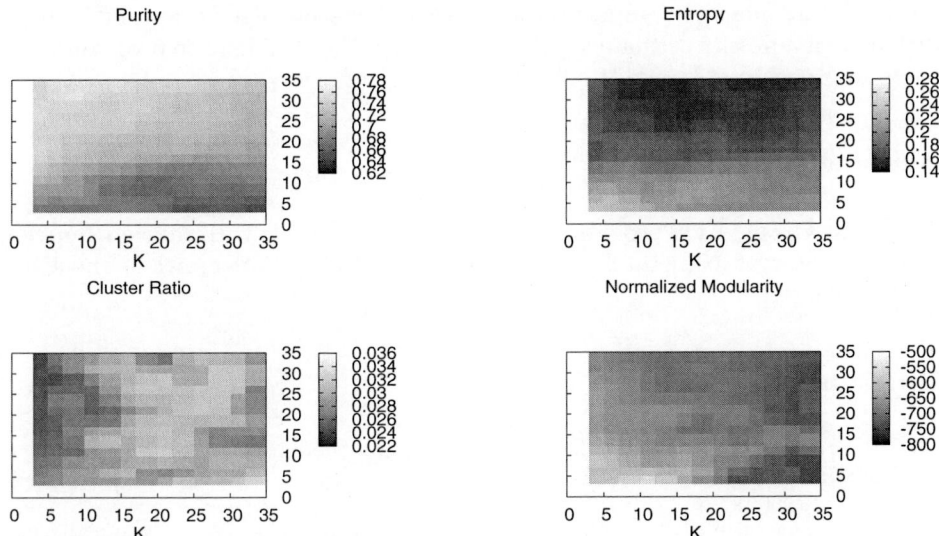

Fig. 5. Relation between dimensions d (vertical), branching k (horizontal) and quality (darkness) for purity (top left), entropy (top right), cluster ratio (bottom left) and normalized modularity (bottom right). The darker the region, the better the clustering quality except for purity where large values denote good output quality.

too large (somewhere between 5 and 10 for this graph) but compute somewhat more eigenvectors.

We also notice a noisy behavior of normalized modularity in Fig. 5, bottom right: counterintuitively and opposite of all other measures, best results are obtained for large branching k regardless of the number of dimensions d or even preferring low dimensions. Due to this unexpected behavior we suspect normalized modularity may not be the appropriate measure for cluster quality in social networks.

4 Conclusion

We gave a k-way hierarchical spectral clustering algorithm with heuristics to balance cluster sizes. We also implemented the heuristics in the recent Divide-and-Merge algorithm [12]. Our algorithm outperformed Divide-and-Merge for clustering the telephone call graph. We also measured the effect of several choices for the input to SVD: we found the weighted Laplacian performing much better than the unweighted counterpart and introduced a neighborhood Jaccard weighting scheme that performs very good for SVD input.

For further work we propose the implementation and comparison of fast SVD approximations and experiments with graphs of even larger scale and in particular of the LiveJournal blogger social network. In its current form our Jaccard weighting scheme requires quadratic space; we believe a fingerprint based approximation such as [40] that can give weight to nonexisting edges will improve the clustering quality. Comparison with other clustering algorithms and in particular with a possible scalable implementation of the semidefinite programming based approaches of Lang [31,32] also remain future work.

Acknowledgment

To Zoltán Gyöngyi for providing us with the host graph with labels from the Open Directory top hierarchy for the UK2007-WEBSPAM crawl of the Ubi-Crawler [9].

References

1. Aiello, W., Chung, F., Lu, L.: A random graph model for massive graphs. In: Proceedings of the 32nd ACM Symposium on Theory of Computing (STOC), pp. 171–180 (2000)
2. Alpert, C.J., Kahng, A.B.: Multiway partitioning via geometric embeddings, orderings, and dynamic programming. IEEE Trans. on CAD of Integrated Circuits and Systems 14(11), 1342–1358 (1995)
3. Alpert, C.J., Kahng, A.B.: Recent directions in netlist partitioning: a survey. Integr. VLSI J. 19(1-2), 1–81 (1995)
4. Alpert, C.J., Yao, S.-Z.: Spectral partitioning: the more eigenvectors, the better. In: DAC 1995: Proceedings of the 32nd ACM/IEEE conference on Design automation, pp. 195–200. ACM Press, New York (1995)
5. Au, W.-H., Chan, K.C.C., Yao, X.: A novel evolutionary data mining algorithm with applications to churn prediction. IEEE Trans. Evolutionary Computation 7(6), 532–545 (2003)
6. Barnes, E.R.: An algorithm for partitioning the nodes of a graph. SIAM Journal on Algebraic and Discrete Methods 3(4), 541–550 (1982)
7. Benczúr, A.A., Csalogány, K., Kurucz, M., Lukács, A., Lukács, L.: Sociodemographic exploration of telecom communities. In: NSF US-Hungarian Workshop on Large Scale Random Graphs Methods for Modeling Mesoscopic Behavior in Biological and Physical Systems (2006)
8. Berry, M.W.: SVDPACK: A Fortran-77 software library for the sparse singular value decomposition. Technical report, University of Tennessee, Knoxville, TN, USA (1992)
9. Boldi, P., Codenotti, B., Santini, M., Vigna, S.: Ubicrawler: A scalable fully distributed web crawler. Software: Practice & Experience 34(8), 721–726 (2004)
10. Chan, P.K., Schlag, M.D.F., Zien, J.Y.: Spectral k-way ratio-cut partitioning and clustering. In: DAC 1993: Proceedings of the 30th international conference on Design automation, pp. 749–754. ACM Press, New York (1993)
11. Cheng, D., Kannan, R., Vempala, S., Wang, G.: On a recursive spectral algorithm for clustering from pairwise similarities. Technical report, MIT LCS Technical Report MIT-LCS-TR-906 (2003)
12. Cheng, D., Vempala, S., Kannan, R., Wang, G.: A divide-and-merge methodology for clustering. In: PODS 2005: Proceedings of the twenty-fourth ACM SIGMOD-SIGACT-SIGART symposium on Principles of database systems, pp. 196–205. ACM Press, New York (2005)

13. Chung, F., Lu, L.: The average distances in random graphs with given expected degrees. Proceedings of the National Academy of Sciences of the United States of America 99(25), 15879–15882 (2002)
14. Chung, F., Lu, L., Vu, V.: Eigenvalues of random power law graphs. Annals of Combinatorics (2003)
15. Chung, F., Lu, L., Vu, V.: Spectra of random graphs with given expected degrees. Proceedings of National Academy of Sciences 100, 6313–6318 (2003)
16. Cormode, G., Indyk, P., Koudas, N., Muthukrishnan, S.: Fast mining of massive tabular data via approximate distance computations. In: ICDE 2002: Proceedings of the 18th International Conference on Data Engineering, p. 605. IEEE Computer Society, Washington (2002)
17. Cox, K.C., Eick, S.G., Wills, G.J., Brachman, R.J.: Brief application description; visual data mining: Recognizing telephone calling fraud. Data Min. Knowl. Discov. 1(2), 225–231 (1997)
18. Derényi, I., Palla, G., Vicsek, T.: Clique percolation in random networks. Physical Review Letters 94, 49–60 (2005)
19. Ding, C.H.Q., He, X., Zha, H.: A spectral method to separate disconnected and nearly-disconnected web graph components. In: KDD 2001: Proceedings of the seventh ACM SIGKDD international conference on Knowledge discovery and data mining, pp. 275–280. ACM Press, New York (2001)
20. Ding, C.H.Q., He, X., Zha, H., Gu, M., Simon, H.D.: A min-max cut algorithm for graph partitioning and data clustering. In: ICDM 2001: Proceedings of the 2001 IEEE International Conference on Data Mining, pp. 107–114. IEEE Computer Society, Washington (2001)
21. Donath, W.E., Hoffman, A.J.: Lower bounds for the partitioning of graphs. IBM Journal of Research and Development 17(5), 420–425 (1973)
22. Drineas, P., Frieze, A., Kannan, R., Vempala, S., Vinay, V.: Clustering large graphs via the singular value decomposition. In: Machine Learning, pp. 9–33 (2004)
23. Fiedler, M.: Algebraic connectivity of graphs. Czechoslovak Mathematical Journal 23(98) (1973)
24. Frieze, A., Kannan, R., Vempala, S.: Fast Monte-Carlo algorithms for finding low rank approximations. In: Proceedings of the 39th IEEE Symposium on Foundations of Computer Science (FOCS), pp. 370–378 (1998)
25. Gorny, E.: Russian livejournal: National specifics in the development of a virtual community. pdf online (May 2004)
26. Gyöngyi, Z., Garcia-Molina, H., Pedersen, J.: Web content categorization using link information. Technical report, Stanford University (2006–2007)
27. Hagen, L.W., Kahng, A.B.: New spectral methods for ratio cut partitioning and clustering. IEEE Trans. on CAD of Integrated Circuits and Systems 11(9), 1074–1085 (1992)
28. Kannan, R., Vempala, S., Vetta, A.: On clusterings — good, bad and spectral. In: IEEE: 2000: ASF, pp. 367–377 (2000)
29. Karypis, G.: CLUTO: A clustering toolkit, release 2.1. Technical Report 02-017, University of Minnesota, Department of Computer Science (2002)
30. Kumar, R., Novak, J., Raghavan, P., Tomkins, A.: Structure and evolution of blogspace. Commun. ACM 47(12), 35–39 (2004)
31. Lang, K.: Finding good nearly balanced cuts in power law graphs. Technical report, Yahoo! Inc. (2004)
32. Lang, K.: Fixing two weaknesses of the spectral method. In: NIPS 2005: Advances in Neural Information Processing Systems, vol. 18, Vancouver, Canada (2005)

33. Malik, J., Belongie, S., Leung, T., Shi, J.: Contour and texture analysis for image segmentation. Int. J. Comput. Vision 43(1), 7–27 (2001)
34. Meila, M., Shi, J.: A random walks view of spectral segmentation. In: AISTATS (2001)
35. Nanavati, A.A., Gurumurthy, S., Das, G., Chakraborty, D., Dasgupta, K., Mukherjea, S., Joshi, A.: On the structural properties of massive telecom graphs: Findings and implications. In: CIKM (2006)
36. Onnela, J.P., Saramaki, J., Hyvonen, J., Szabo, G., Lazer, D., Kaski, K., Kertesz, J., Barabasi, A.L.: Structure and tie strengths in mobile communication networks (October 2006)
37. Open Directory Project (ODP), http://www.dmoz.org
38. Richardson, M., Domingos, P.: Mining knowledge-sharing sites for viral marketing. In: KDD 2002: Proceedings of the eighth ACM SIGKDD international conference on Knowledge discovery and data mining, pp. 61–70. ACM Press, New York (2002)
39. Sarlós, T.: Improved approximation algorithms for large matrices via random projections. In: Proceedings of the 47th IEEE Symposium on Foundations of Computer Science (FOCS) (2006)
40. Sarlós, T., Benczúr, A.A., Csalogány, K., Fogaras, D., Rácz, B.: To randomize or not to randomize: Space optimal summaries for hyperlink analysis. In: Proceedings of the 15th International World Wide Web Conference (WWW), pp. 297–306 (2006)
41. Shi, J., Malik, J.: Normalized cuts and image segmentation. IEEE Transactions on Pattern Analysis and Machine Intelligence (PAMI) (2000)
42. Shiga, M., Takigawa, I., Mamitsuka, H.: A spectral clustering approach to optimally combining numerical vectors with a modular network. In: KDD 2007: Proceedings of the 13th ACM SIGKDD international conference on Knowledge discovery and data mining, pp. 647–656. ACM Press, New York (2007)
43. von Luxburg, U., Bousquet, O., Belkin, M.: Limits of spectral clustering, pp. 857–864. MIT Press, Cambridge (2005)
44. Wei, C.-P., Chiu, I.-T.: Turning telecommunications call details to churn prediction: a data mining approach. Expert Syst. Appl. 23(2), 103–112 (2002)
45. Weiss, Y.: Segmentation using eigenvectors: A unifying view. In: ICCV (2), pp. 975–982 (1999)
46. Wills, G.J.: NicheWorks — interactive visualization of very large graphs. Journal of Computational and Graphical Statistics 8(2), 190–212 (1999)
47. Zakharov, P.: Structure of livejournal social network. In: Proceedings of SPIE, vol. 6601, Noise and Stochastics in Complex Systems and Finance (2007)
48. Zha, H., He, X., Ding, C.H.Q., Gu, M., Simon, H.D.: Spectral relaxation for k-means clustering. In: Dietterich, T.G., Becker, S., Ghahramani, Z. (eds.) NIPS, pp. 1057–1064. MIT Press, Cambridge (2001)

Looking for Great Ideas: Analyzing the Innovation Jam*

Wojciech Gryc, Mary Helander, Rick Lawrence, Yan Liu,
Claudia Perlich, Chandan Reddy, and Saharon Rosset

IBM T.J. Watson Research Center
P.O. Box 218
Yorktown Heights, NY 10598

Abstract. In 2006, IBM hosted the Innovation Jam with the objective of identifying innovative and promising "Big Ideas" through a moderated on-line discussion among IBM worldwide employees and external contributors. We describe the data available and investigate several analytical approaches to address the challenge of understanding "how innovation happens". Specifically, we examine whether it is possible to identify characteristics of such discussions that are more likely to lead to innovative ideas as identified by the Jam organizers. We demonstrate the social network structure of data and its time dependence, and discuss the results of both supervised and unsupervised learning applied to this data.

1 Introduction

Social network analysis is the mapping and measuring of relationships and flows between people, groups, organizations or other information/knowledge processing entities. There has been tremendous work on the social network study over the past century [14,2,11]. Nowadays commoditization and globalization are dominant themes having a major impact on business execution. As a result, large companies are focusing extensively on innovation as a significant driver of the new ideas necessary to remain competitive in this evolving business climate. Of course, the broader issue is how does a company foster innovation, and specifically how do we identify, extend, and capitalize on the new ideas that are created?

With the wide use of worldwide web, people are provided a much more convenient and quick means for communication so that much larger and richer "virtual" social networks are formed, such as "MySpace", "Facebook" and "LinkedIn". One type of the common virtual world is the forum, where people can discuss topics of interest online at any time and any place. Such virtual worlds increasingly arise even within corporate environments [12].

---

* This work is base on an earlier work: Looking for Great Ideas - Analyzing the Innovation Jam in WebKDD/SNA-KDD '07: Proceedings of the 9th WebKDD and 1st SNA-KDD 2007 workshop on Web mining and social network analysis ©ACM, 2007. http://doi.acm.org/10.1145/1348549.1348557

H. Zhang et al. (Eds.): WebKDD/SNA-KDD 2007, LNCS 5439, pp. 21–39, 2009.
© Springer-Verlag Berlin Heidelberg 2009

IBM recently introduced an online information forum or "Innovation Jam" [9,13] where employees (and, in some cases, external participants) are encouraged to share their ideas on pre-selected topics of broad interest. Analysis of the information collected in such forums requires a number of advanced data processing steps including extraction of dominant, recurring themes and ultimately characterization of the degree of innovation represented by the various discussion threads created in the forum. Topic identification [7] poses a significant challenge in any unstructured forum like a blog, but it is perhaps less of an issue in the Jam data due to the apriori topic suggestions. As described in Section 3, the Jam consisted of two successive phases, followed by a selection of highly promising ideas based on these discussions. This multi-stage format provides a rich set of data for investigation of how ideas evolve via moderated discussion. Of particular interest is whether we can detect characteristics of those discussion threads that can be linked to an idea ultimately being selected as a promising initiative. To the extent that selected ideas reflect some indication of "innovation," we have some basis for examining which aspects of a discussion thread may lead to innovation. This paper summarizes our efforts to characterize successful threads in terms of features drawn from both the thread content as well as information like the organizational divergsity of the participants in such threads.

The paper is organized as follows. In Section 2, we discuss some metrics that have been used to characterize the structure of social networks formed via other kinds of discussion groups. Section 3 describes the specifics of the IBM Innovation Jam and the collected data. Section 4 summarizes some key aspects of the dynamics of the Jam interactions. Finally, Sections 5 and 6 describe respectively the unsupervised and supervised learning approaches we have applied to this data.

2   Related Work

The Innovation Jam is a unique implementation of threaded discussions, whereby participants create topics and explicitly reply to each other using a "reply to" button. As such, discussions are hierarchical, with a clear flow of messages from the initial parent post to subsequent ideas and thoughts. Unlike more unstructured discussions, the person's response is then only associated with the message he or she chose to reply to. Through studies focusing on usenet groups and online forums, a great deal is known about the types of social structures that may develop within such communities. Indeed, it has been observed that depending on discussion topics, group membership and other features of a usenet group or forum, differing social structures may develop.

For instance, Turner et al. [15] show that newsgroups vary a great deal based on the amount of time individuals spend on a single visit, and the longevity of individuals' involvement in the groups themselves. The authors specifically break down the social roles of participants into types like the "questioner", "answer person", "spammer", and "conversationalist", among others. Depending on the participants, such groups can then develop different forms of participation and norms. For example, a questioner may post once in a group and never again, while an answer person may respond to a large number of questions, but never participate in continuing dialogues. A group of conversationalists can be characterized by a large number of posts and replies to each other.

While the social network analysis of Innovation Jam participants is beyond the scope of the paper, it is important to note that the social structure of the Jam has an effect on its outcomes. Newsgroups or forums can develop their own internal social structures. Groups can also be seen as those that develop unique linguistic or social traits [6]. For example, groups may begin to use new words or acronyms, lending support to the idea that predicting convergence in discussions or brainstorms is possible.

As described in the following section, the Innovation Jam format has been used extensively within IBM. The concept has also been extended to external events like Habitat Jam, at the World Urban Forum (www.wuf3-fum3.ca) in 2006. The brainstorming format within a limited time frame is recognized as a useful approach to dispersed and large-scale collaboration. Thus, it is useful to analyze the discussions, and explore how people reached the resultant "big deas" of the Jam.

3 Innovation Jam Background

In 2001, IBM introduced the Jam concept through a social computing experiment to engage large portions of its global workforce in a web-based, moderated brainstorming exercise over three days [8]. What became known as the "World Jam" was eventually followed by six additional internal, corporate-wide Jams, drawing employees into discussions about everything from management to company values. In early 2006, IBM announced that it would again use the Jam concept for an eighth time - this time, for facilitating innovation among the masses, and also including participants from external organizations and IBM employee family members.

Key to the design of the Jam's large scale collaborative brainstorming methodology was the identification of seed areas. Before the event launch, teams were formed to brainstorm general areas and to discuss the design and implementation details. Four general areas, called *"Forums,"* were identified:

- **Going Places** - Transforming travel, transportation, recreation and entertainment
- **Finance & Commerce** - The changing nature of global business and commerce
- **Staying Healthy** - The science and business of well-being
- **A Better Planet** - Balancing economic and environmental priorities

Factors that determined the selection of seed areas included the opinions of IBM technical leaders, the technical relevance to IBM's business strategies, as well as the overall relevance to general societal and global economic challenges.

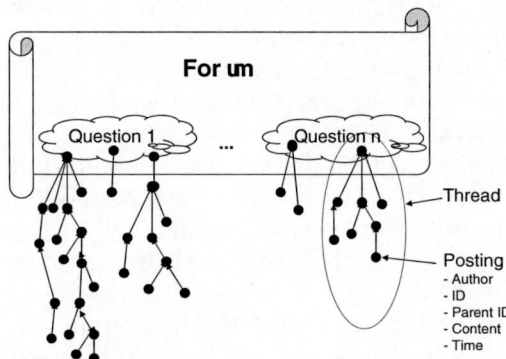

Fig. 1. Relationship between postings, threads, questions and forums in both Jam phases

3.1   The Innovation Jam Process

IBM's Innovation Jam was designed to take part over two phases. Phase 1 took place July 24-27, 2006 and primarily focused on idea creation and development. Unlike previous IBM Jams where preparation was not necessary, the Jam required familiarization with emerging technologies which were described in online materials made available to participants prior to the event.

Individual contributions to the Jam came in the form of "*postings*," or messages in reply to other contributors and to questions posed under a moderated topic area. As shown in Figure 1, groups of such postings are defined as "*threads*".

For five weeks following Phase 1 of the Innovation Jam, a multi-discipline, international cross-IBM team analyzed more than 37,000 Phase 1 posts to identify the most promising suggestions, resulting in 31 identified topics or "*big ideas*" as listed in Table 4. Taking the raw ideas from Phase 1 and transforming them into real products, solutions and partnerships to benefit business and society was the focus of Innovation Jam Phase 2, September 12-14, 2006, which involved more focused sessions where participants refined ideas.

In light of the discussion in Jam Phase 2, internal teams drawing on a broad range of subject-matter expertise continued to discuss and shape the key emerging ideas, evaluating them in terms of their technological innovation as well as their potential impact on society and business. Based on these discussions, a final list of ten leading ideas or "*finalists*" were identified to receive funding for development over the next two years. The ten finalists were:

1. **3-D Internet:** Establish the 3-D Internet as a seamless, standards-based, enterprise-ready environment for global commerce and business.
2. **Big Green innovations:** Enter new markets by applying IBM expertise to emerging environmental challenges and opportunities.
3. **Branchless Banking:** Profitably provide basic financial services to populations that don't currently have access to banking.

4. **Digital Me:** Provide a secure and user-friendly way to seamlessly manage all aspects of my digital life - photos, videos, music, documents, health records, financial data, etc. - from any digital device.
5. **Electronic Health Record System:** Create a standards-based infrastructure to support automatic updating of - and pervasive access to healthcare records.
6. **Smart Healthcare Payment System:** Transform payment and management systems in healthcare system
7. **Integrated Mass Transit Information System:** Pursue new methods to ease congestion and facilitate better flow of people, vehicles and goods within major metropolitan areas.
8. **Intelligent Utility Network:** Increase the reliability and manageability of the world's power grids.
9. **Real-Time Translation Services:** Enable innovative business designs for global integration by removing barriers to effective communication, collaboration and expansion of commerce.
10. **Simplified Business Engines:** Deliver the "Tunes" of business applications.

While recognizing that significant human processing took place in the course of evaluating Jam data, our goal is to see if we can identify factors that would have been predictive of the Jam finalists, perhaps suggesting ways to help make processes for future Jams less manually intensive.

3.2   Overview of the Jam Characteristics

As mentioned above, the Innovation Jam was conducted in two phases that were separated by a period of less than 2 montha. Table 1 summarizes some of the basic statistics of these two phases. In both phases, all the threads belonged to one of the following four forums: (1) Going Places, (2) Staying Healthy, (3) A Better Planet and (4) Finance and Commerce

Figure 2 gives the percentage of messages in each of the above mentioned forums during Phase 1 and Phase 2. We can see that topics related to "Going Places" received relatively more attention during Phase 2. Percentage of contributors who responded more than 1-20 times during both phases is shown in Fig. 3. Considering the fact that the numbers of contributors are 13366 in Phase

Table 1. Summary statistics for the two phases conducted in Innovation Jam

Summary Statistics	Phase 1	Phase 2
No. of Messages	37037	8661
No. of Contributors	13366	3640
No. of Threads	8674	254
No. of Threads with no response	5689	0
No. of Threads with ≤10 responses	2673	60
No. of Threads with ≥100 responses	56	12

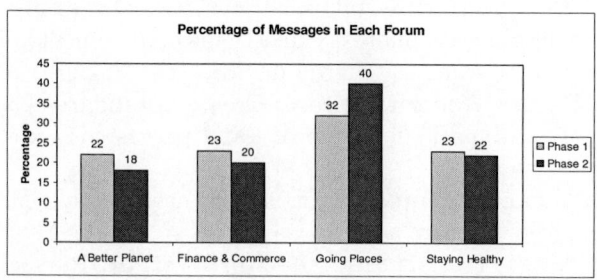

Fig. 2. Percentage of messages in each forum for Phase 1 and Phase 2

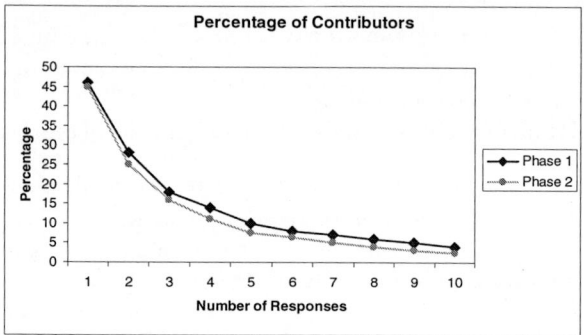

Fig. 3. Percentage of contributors who responded more than 1-20 times during Phase 1 and Phase 2

1 and 3640 in Phase 2, it is interesting to note that these percentages are very similar for both phases. For example, percentage of contributors who responded at least 3 times is 18% for Phase 1 and 16% for Phase 2.

3.3   Sources of Data

In the case of the IBM Innovation Jam, we have access to unique and highly-diverse sources of high quality data to be used in addressing our learning challenge. We now briefly review the data sources and types we have available. In the succeeding sections we will describe the data itself and our analytical approaches.

Since a prevalent hypothesis is that a major advantage of the Jam is that it brings together people from different parts of IBM, and different geographical locations, who would otherwise be unlikely to interact — and that such interactions between diverse groups are likely to lead to new insights and innovations — this data is of particular interest in our analysis. These data sources are:

1. **The text of the threads itself.** From analyzing the text we can find similarity between threads, understand how tight the discussion in each thread was, identify the keywords differentiating between threads.

2. **The social network structure of threads and the whole Jam.** Within each thread, we can analyze the structure of the discussion, and collect statistics such as how many "leaves" (postings with no response) there were, how deep is the typical discussion in the thread, etc. Since we have unique identifiers for all contributors, we can also analyze the connection between threads through common contributors, the variety of contributors in each thread (e.g, messages per poster).

3. **The organizational relationships between the contributors.** Since the vast majority of contributors were IBM employees, we can make use of the online directory of worldwide IBM employees (known as Blue Pages), to capture the organizational and hierarchical relationships between the contributors in each thread, in each *Big Idea*, etc.

4   Social Network and Dynamics in the Jam Interactions

Let us take a closer look at the social aspect of the Jam domain and in particular how the network of interactions between contributors evolves over time. Figure 4 shows the number of postings per hour over the 3 days of the Phase 1. The plot shows after an initial spike within the first two hours clear seasonality of a 24 hour rhythm. The hourly count of contributions remains fairly stable over the 3 day period. In the sequel we will consider the social network of contributors where every node is a contributor and a directed link from person A to person B is present if A directly responded to B. We can extract this relationship from the posting identifiers and the provided identifier of the parent posting. Here, social behavior is modeled as a network in an approach similar to [1].

The resulting social Jam network is shown for a number of points in time (2 hours, 3 hours, 4 hours and 10 hours after start of Jam) in Figure 5. The graph

Fig. 4. Number of postings over time during Jam Phase 1

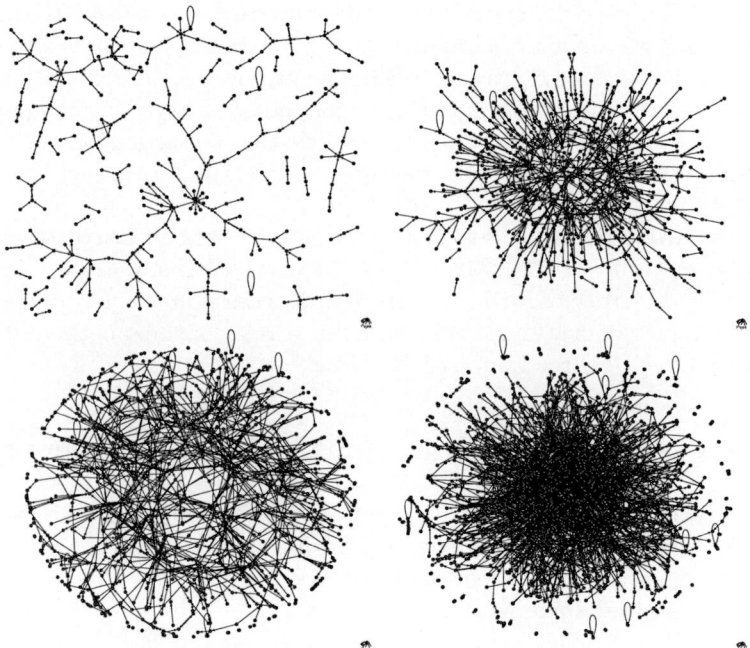

Fig. 5. Evolvement of the social Jam network for 2, 3, 4 and 10 hours after the Phase 1 start

layouts are generated using Pajek [3]. The initial network has after 2 hours still a number of independent components that most likely reflect the thread structure of the Jam. However, already after 3 hours the Jam population is fairly interconnected and only a few very small independent components remain. This trend continues until after 10 hours the network structure compacts into a tight ball with a number of small peripheral components. Given the linear trend in the population and the rather constant rate of postings, the overall density (number of present links over number of possible links) of the social network is exponentially decreasing.

Figure 6 shows a histogram of the organizational distances between IBM contributors who posted replies to other IBM contributors. (Posts connecting IBM contributors and external people were not considered.) Here, organizational distance between two contributors is determined by traversing the reporting structure until a common person is found, and counting the number of nodes (people) encountered in the complete traversal. This distance was incremented by 2 to bridge geographies when reporting trees were not connected. The average distance of 11.6, and the wide distribution of distances, suggests that the Innovation Jam was successful as a forum for creating interactions between people who otherwise might have been unlikely to interact.

The observed network structure suggests that the individual Jam contributors are not focused on a single thread but rather seem to 'browse' between threads

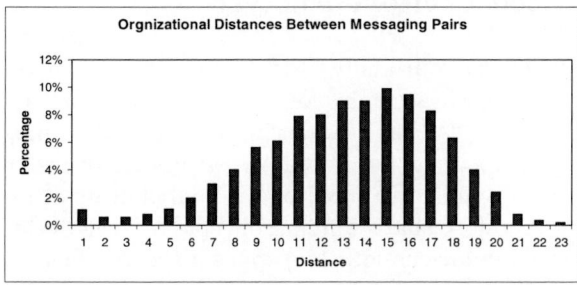

Fig. 6. Organizational distances between IBM contributors who posted replies to other IBM contributors

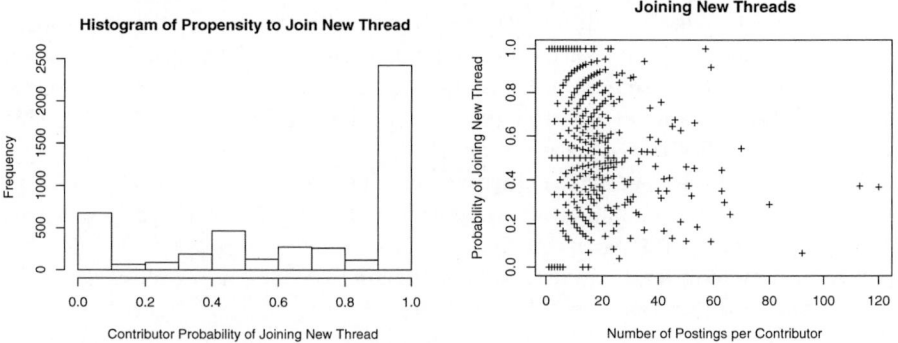

Fig. 7. Histogram and scatterplot of the propensity of contributors to post to a thread they have not posted to before rather than positing to a thread they did contribute to in the past

and topics. If individual contributors were contributing to a single thread we would expect the network to show a number of loosely connected islands (corresponding to threads) with high interconnectivity. As a first consideration we estimate the average probability of a repeating contributor to post to a new thread, where new is defined as a thread to which he has never posted. We define this probability for a given contributes as the number of contributed posts minus one (the first posting is by definition to a thread that the contributer has not posted in before) divided by the number of different threads in which he posted.

And indeed, this probability is surprisingly high at 62%. The histogram in Figure 7 shows that a large number of Jam contributors ventured into multiple threads. The large spike around probability 1 is caused by contributors with only 2 postings in two different threads. However, the scatter plot reveals that there is no clear functional dependence between the number of postings of a contributor and his probability of contributing to multiple threads.

5 Unsupervised Content Analysis

The high-level challenge of our analysis of the Innovation Jam data is to identify the keys to success of such an endeavor, in particular, what are the characteristics of discussion threads that lead to innovative and promising ideas? The major differences between the Jam data and a typical forum are: a) the topics are more concentrated and controlled; b) the contributors are mostly from one organization, and therefore share similar concepts on basic values and what are the "great" ideas; c) the discussion time spans a shorter time.

As in every learning problem, there are two general approaches that can be taken to address this challenge:

- **The supervised learning approach.** If we could go back and label discussion threads as *successful* or *unsuccessful*, we could then investigate and characterize the features differentiating between the two classes, and hypothesize that these are the features that lead to success. As we discuss below, we have utilized the selection of big ideas from the Jam as finalists for funding for labeling, and attempted to correlate the various features with this selection, with limited success so far.
- **The unsupervised learning approach.** The idea here is to concentrate our effort on characterizing and categorizing the discussion threads in terms of their typical profiles, or groups of distinct typical profiles. While this analysis may not lead directly to conclusions on which profiles represent *successful* Jam threads, it can be an important step towards hypothesis generation about success. Furthermore, it can be used as an input to discussions with experts and to design of experiments to test the success of the different thread types in generating innovation. We describe the promising results of unsupervised learning on Jam text features below.

We discuss the unsupervised approach in this section, and defer the discussion of supervised techniques to Section 6.

5.1 Data Preprocessing

To analyze the content of the jam posting, we preprocess the text data and convert them into vectors using bag-of-words representation. More specifically, we put all the postings within one thread together and treat them as one big document. To keep the data clean, we remove all the threads with less than two postings, which results in 1095 threads in Phase 1 and 244 threads in Phase 2. Next, we remove stop words, do stemming, and apply the frequency-based feature selection, i.e. removing the most frequent words and those appearing less than 2 times in the whole collection. These processes results in a vocabulary of 10945. Then we convert the thread-level documents into the feature vectors using the "ltc" TF-IDF term weighting [4].

5.2   Clustering Algorithm

Our objective of the unsupervised analysis is to find out what are the overlapping topics in Phase 1 and Phase 2, i.e. the topics that discussed in Phase 1 have been picked up by Phase 2, which can be seen as a potential indicator of "finalists" of ideas for funding. Therefore when we cluster the threads from Phase 1 and Phase 2, an optimal case is that we can find three types of clusters: (1) the clusters that mostly consist of threads in Phase 1 (2) those mostly composed of threads in Phase 2; and (3) the clusters with the threads in both phases, which help us examine if they are the potential finalists for funding. Several clustering algorithms have been investigated, including K-means, hierarchical clustering, bisecting K-means and so on [10]. The results from different clustering algorithms are similar and therefore we only discuss the ones using the complete-linkage agglomerate clustering algorithm. For implementation,we use the open source software CLUTO[1].

5.3   Clustering Results

As discussed above, we use the document clustering algorithms to analyze the threads in Phase 1 and Phase 2. In the experiment, we preset the number of clusters to 100. Several interesting observations can be made by examining the clustering results: (1) Phase 1 to Phase 2: since we are interested in finding out the overlapping topics between the threads in Phase 1 and those in Phase 2, we plot the histogram on the number of threads from Phase 2 in each cluster in Figure 8. From the results, we can see that the majority of the clusters (around 70%) only contain the threads in Phase 1, which indicate that the topics in phase 2 are only a subset of those in Phase 1 and there are no new topics in Phase 2. This agrees well with the process of the Jam, i.e. a subset of the topics discussed in Phase 1 are selected and used as discussion seed in Phase 2. (2) Phase 2 to Finalist ideas: we further examine the topic overlapped between the threads in Phase 2 and those selected as successful finalist ideas by going through the clusters with the most threads from Phase 2. From Table 2, we can see an impressively direct mapping from the top-ranked clusters (by the number of threads from Phase 2) to the finalist. For example, the cluster with the largest number of threads from Phase 2 is shown in the first line. It seems to concentrate on the topics about "patients", "doctors" and "healthcare", which agrees well the main theme in one of the finalist ideas, i.e. "Electronic Health Record System". Another example is the cluster devoted to the idea of "Digital Me". Its descriptive words are "dvd", "music", "photo" and so on, which clearly reflects the theme about providing a secure and user-friendly way to seamlessly manage photos, videos, music and so on.

5.4   Topic Tracking

In this section, we explore the evolution of discussion concentration over time. In particular, we are interested in answering the questions: What are the possible

---

[1] http://glaros.dtc.umn.edu/gkhome/views/cluto

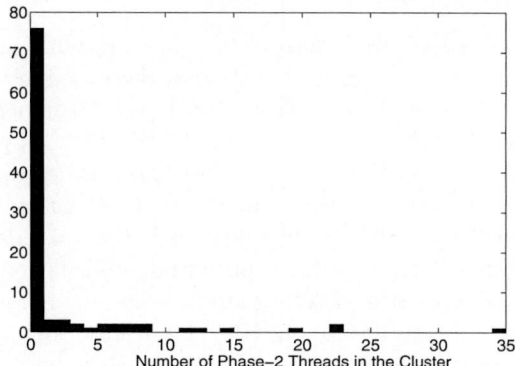

Fig. 8. Histogram of the number of Phase 2 threads in the 100 clusters

characteristics of a focused discussion, and are there any patterns as to when and how the focused discussion can happen? The three largest threads were selected for analysis, with the first discussing "Sustainable energy systems", the other on "Digital entertainment" and the third on "Global travel" with 622, 405 and 776 posts respectively.

To determine whether a series of posts have a focused discussion or not, we compare the content similarity as well as the number of unique contributors. We preprocess the texts in each post using the bag-of-words representation described before, and then calculate the averaged cosine similarity between 10 adjacent posts in a time-based consecutive order. In addition, we calculate the statistics of the number of unique contributors in a way that penalizes discussions involving very few contributors (to avoid random chatting between two persons). Figure 9 shows the discussion concentration with content similarity and the number of unique contributors. There seems to be a general trend common in all the three threads, that is, at the beginning of the Jam there are many posts on multiple topics, then these topics evolve as time goes, and finally one or more of them lead to a focused discussion. Usually these effective discussions appear at around 20 to 30 hours after the Jam starts.

To further examine the effectiveness of our concentration measurement, we show an example of 5 posts with the highest averaged cosine-similarity score and the most unique contributors in the "digital entertainment" thread. These posts occur at around 27 hours after the jam starts as identified in Figure 9).

1. "... Going to the movies is a social experience. It is about enjoying the whole experience: large screen, popcorn, people ..."
2. "... if you want to experience that you might want to go to disneyland to see/feel 'honey I shrunk the audience"'
3. "The possible future development in entertainment will be the digital eye glasses with embedded intelligence in form of digital eye-glasses. The advantages for users would be: the screen size and the 'feeling' to be inside the action ..."

Table 2. The mapping from the clusters with the most threads in Phase 2 to the finalist ideas. P1 and P2 are the number of threads in the cluster from Phase 1 and from Phase 2 respectively.

Finalist Ideas for Funding	P1	P2	Descriptive Stemmed Words
Electronic Health Record System	49	35	patient, doctor, healthcar, diagnosi, hospit, medic, prescript, medicin, treatment, drug, pharmaci, nurs, physician, clinic, blood, prescrib, phr, diagnost, diseas, health
Digital Me	26	23	scrapbook, music, dvd, song, karaok, checker, entertain, movi, album, content, artist, photo, video, media, tivo, piraci, theater, audio, cinema
Simplified Business Engines	26	23	smb, isv, back-offic, eclips, sap, mashup, business-in-a-box, invoic, erp, mgt, oracl, app, salesforc, saa, host, procur, payrol, mash, crm
Integrated Mass Transit Information System	59	20	bus, congest, passeng, traffic, railwai, commut, rout, lane, destin, transit, journei, rail, road, vehicl, rider, highwai, gp, driver, transport
Big Green innovations	27	13	desalin, water, rainwat, river, lawn, irrig, rain, filtrat, purifi, potabl, osmosi, contamin, purif, drink, nanotub, salt, pipe, rainfal, agricultur
3-D Internet	22	12	password, biometr, debit, authent, fingerprint, wallet, finger, pin, card, transact, atm, merchant, reader, cellular, googlepag, wysiwsm, byte, userid, encrypt
Intelligent Utility Network	23	9	iun, applianc, peak, thermostat, quickbook, grid, outag, iug, shut, holist, hvac, meter, heater, household, heat, resours, kwh, watt, electr, fridg
Branchless Banking	11	9	branchless, banker, ipo, bank, cr, branch, deposit, clinet, cv, atm, loan, lender, moeni, withdraw, teller, mobileatm, transact, wei, currenc, grameen
Real-Time Translation Services	33	5	mastor, speech-to-speech, speech, languag, english, nativ, babelfish, translat, troop, multi-lingu, doctor-pati, cn, lanaguag, inno, speak, arab, chines, barrier, multilingu

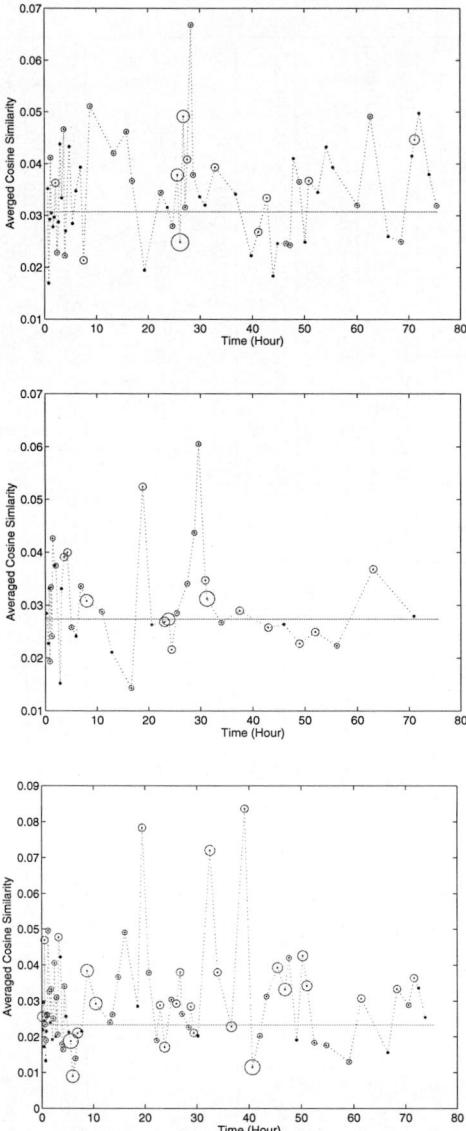

Fig. 9. Plot of averaged similarity between 10 posts in most common threads; the size of the circle indicate the inverse of number of unique contributors (the bigger the size, the lower the number of unique contributors). The green line is the cosine similarity between a random pair of posts within the thread (the score is usually between 0.02 and 0.03). Topic discussed include, Top: Sustainable energy systems, Middle: Digital entertainment, Bottom: Global travel.

4. " ... Really big screens accompanied by great sound and special lighting effects would really enhance the experience and make it different from renting a movie at home ..."

5. "It would be nice if multiple movies could play on the same screen and the audience would wear special glasses so they could see only the movie they payed for. This would reduce the need for multiple theater auditoriums in a single location."

We can see that the discussion is significantly focused on multiple ways of improving the theater experience.

6   Finding Great Ideas: Supervised Analysis

Several features are extracted from the Jam data. More emphasis is given to the Phase 2 interactions because of the fact that the *finalists* were selected from Phase 2 of the Innovation Jam. A total of eighteen features (three different categories) were obtained:

1. **Topological Features:** Features T1-T8 described in Table 3 correspond to topological features. These features will give some basic intuition about the Phase 2 of the Innovation Jam. It contains information regarding the topology of the messaging including number of messages, number of contributors, number of questions in a given idea, number of responses for each question and so on. Column T8 corresponds to the interconnection of contributors between these ideas. It gives the number of times that the contributors of a given idea participated in other ideas. The contributors are weighted based on their contribution in the given idea.

2. **Contextual Features:** Features C1-C5 described in Table 3 correspond to contextual features. These features are computed based on the bag-of-words representation of all the messages belonging to a single thread. The pairwise cosine similarity measure is computed between all possible pairs of threads with more than one message in a particular big idea. Some basic statistics like the mean, standard deviation, maximum and minimum of these scores are considered as features.

3. **Organizational Features:** Features O1-O5 described in Table 3 correspond to organizational features. Basically, organizational distance between two contributors can be computed by traversing a 'management tree' where each node corresponds to a person and its parent node corresponds to the person to whom he reports to. The distance between two contributors can be obtained by *climbing up* each of the trees until a common person is found[5]. Sometimes, two contributors might not have any common personnel in the

---

[3] Excluding the questions with less than 10 responses.

[4] Threads containing more than one message.

[5] For only those contributors whose organizational information was available.

[5] For few contributors, it was difficult to obtain the organizational hierarchy information. These cases were eliminated during the computation.

Table 3. Description of 18 different features used in the analysis of Innovation Jam

Index	Description of the Feature	t-test	K-S test	M-W test
T1	Total Number of messages for a particular big idea.	0.58	0.99	0.67
T2	Total Number of messages which didn't receive any further response.	0.61	0.97	0.60
T3	Total Number of contributors.	0.92	0.94	0.95
T4	Forum Number.	0.86	1	0.90
T5	Total Number of questions asked in that particular idea.	0.70	0.91	0.71
T6	Mean of the number of messages for all questions [2].	0.96	0.69	0.82
T7	Standard deviation of the number of messages for all questions [2].	0.53	0.90	0.66
T8	Weighted number of overlapping contributors involved in other big ideas.	0.91	0.88	1
C1	Mean of the pairwise cosine similarity scores between the threads [3].	0.31	0.70	0.34
C2	Standard deviation of the pairwise scores between the threads [3].	0.40	0.29	0.28
C3	Total number of pairwise scores between all threads.	0.52	0.85	0.46
C4	Maximum pairwise score between the threads.	0.38	0.91	0.90
C5	Minimum pairwise score between the threads.	0.94	0.84	0.79
O1	Average pairwise distance between the contributors within a big idea [4].	0.62	0.66	0.54
O2	Standard deviation of the pairwise distances between the contributors [4].	0.91	0.94	0.97
O3	Total number of pairwise distances between all the contributors involved.	0.93	0.90	0.98
O4	Maximum pairwise distance between the contributors.	0.64	0.85	0.59
O5	Minimum pairwise distance between the contributors.	0.046	0.29	0.046

Table 4. Summary of 31 big idea names obtained from the analysis of Phase 2 and label indicating whether they were under the finalists selected for funding

Big Idea	Funded
Rail Travel for the 21st Century	0
Managed Personal Content Storage	1
Advanced Safecars	0
Health Record Banks	1
The Truly Mobile Office	0
Remote Healthlink	0
Real-Time Emergency Translation	1
Practical Solar Power Systems	0
Big Green Services	1
Cellular Wallets	0
Biometric Intelligent Passport	0
Small Business Building Blocks	0
Advance Traffic Insight	0
3-D Internet	1
Branchless Banking for the Masses	1
e-Ceipts	0
Digital Entertainment Supply Chains	0
Smart Hospitals	0
Business-in-a-box	1
Retail Healthcare Solutions	0
Digital Memory Saver	0
Intelligent Utility Grids	1
Cool Blue Data Centers	0
Water Filtration Using Carbon Nanotubes	0
Predictive Water Management	0
Sustainable Healthcare in Emerging Economies	0
Bite-Sized Services For Globalizing SMBs	0
Integrated Mass Transit Information Service	1
Smart-eyes, Smart-insights	0
Smart Healthcare Payment Systems	1
Advanced Energy Modelling and Discovery	0

reporting structure. In those cases, both the lengths of the reporting structure for the two contributors are added and the total is incremented by 2 (considering the fact that people in the topmost position in the ladder are somehow connected by another imaginary layer). Again, some basic statistics are computed as described above.

The values of these eighteen features are computed for all the 31 big ideas (Table 4). We also associate a *label* field with each big idea, indicating whether or not it was chosen as a "finalist" for funding. Hence, we can treat this as a supervised learning problem and we can use the labeling to identify the most informative features.

Testing the features for association with selection for funding. We investigated the correlation between our 18 features and the response variable — whether or not each "big idea" was selected as a finalist for funding. We applied a parametric t-test, and two non-parametric tests (Kolmogorov-Smirnov and Mann-Whitney, [5]) to test the hypothesis of a difference in the distributions P(feature|selected) and P(feature|not selected) for each of the 18 features. The results (Table 3) demonstrate that there is no evidence that any of the features carries significant information about the selection process. The last feature, *Minimum pairwise distance between the contributors*, results in a p-value that is smaller than 0.05 for a couple of tests, but given the amount of multiple comparisons we are doing, this can by no means be taken as evidence of real association. Thus we can conclude that our 18 features fail to capture the "essence" of the Jam as it pertains to the finalist funding decisions. Discovering and formalizing this essence remains a topic for future work.

7   Conclusion

Our broad objective in this initial work is to apply machine-learning-based techniques to data obtained from moderated, online forums like Innovation Jam, with the purpose of identifying aspects of these discussions that lead to innovative ideas. This is a particularly challenging task, and we have applied both supervised and unsupervised approaches to the Jam data, which includes labeling of the most innovative ideas based on human-expert insight. Examination of a range of features drawn from analysis of the topology of the discussion, the context of the discussion, and the organizational diversity of the participants did not yield strong statistical evidence concerning how innovative ideas evolve.

Although this supervised study was not as successful as hoped, this work has shown that the Jam data does exhibit some ordering both in terms of the social structure and discussions themselves. It has been observed that within short time frames (i.e. within minutes), discussions between individuals can be observed, and that over time, (i.e. hours and days) general discussions tend to become more focused and specific.

The Innovation Jam was a unique implementation of a threaded discussion due to its corporate focus, short time frame, and use of moderators. It is encouraging to see that even in such a short period, collaboration can be observed and people can begin working together to generate novel ideas. Much work is left in extending our use of the different data types in both supervised and unsupervised learning, and in identifying the key characteristics — or combination of characteristics — that lead to success.

Acknowledgments

We would like to thank Cathy Lasser, Samer Takriti, Jade Nguyen Strattner for general discussions and insights about the IBM Innovation; Noel Burke,

Selena Thomas for help finding information on the IBM Innovation Jam process; William Tuskie and Scott Sprangler for providing the Jam data; Cathy Lasser, David Yaun for their critical review and discussion of our conclusions; and John Thomas and Kate Ehrlich for insights regarding social networks, and socio-technical systems.

References

1. Agrawal, R., Rajagopalan, S., Srikant, R., Xu, Y.: Mining newsgroups using networks arising from social behavior. In: Proceedings of the 12th International Conference on World Wide Web, pp. 529–535 (2003)
2. Albert, R., Barabási, A.-L.: Statistical mechanics of complex networks. Reviews of Modern Physics 74(1), 47–97 (2002)
3. Batagelj, V., Mrvar, A.: Pajek - program for large network analysis. Connections 2(21), 47–57 (1998)
4. Buckley, C., Salton, G., Allan, J.: The effect of adding relevance information in a relevance feedback environment. In: Proceedings of the Seventeenth Annual International ACM-SIGIR Conference on Research and Development in Information Retrieval. Springer, Heidelberg (1994)
5. Conover, W.J.: Practical nonparametric statistics. John Wiley and Sons, New York (1971)
6. Golder, S.A., Donath, J.: Social roles in electronic communities. Internet Research 5, 19–22 (2004)
7. Gruhl, D., Guha, R., de Liben-Nowell, D., Tomkins, A.: Information diffusion through blogspace. In: Proceedings of the 13th International Conference on World Wide Web(WWW), pp. 491–501 (2004)
8. Halverson, C., Newswanger, J., Erickson, T., Wolf, T., Kellogg, W.A., Laff, M., Malkin, P.: World jam: Supporting talk among 50,000+. In: Proceedings of the European Conference on Computer-Supported Cooperative Work (ECSCW) (2001)
9. Hempel, J.: Big blue brainstorm. Business Week 32 (August 2006)
10. Jain, A.K., Dubes, R.C.: Algorithms for clustering data. Prentice-Hall, Inc., Upper Saddle River (1988)
11. Kleinberg, J.: The Small-World Phenomenon: An Algorithmic Perspective. In: Proceedings of the 32nd ACM Symposium on Theory of Computing (2000)
12. Kolari, P., Finin, T., Lyons, K., Yeshaa, Y., Yesha, Y., Perelgut, S., Hawkins, J.: On the structure, properties and utility of internal corporate blogs. In: International Conference on Weblogs and Social Media (2007)
13. Lasser, C.: Discovering innovation. In: IEEE International Conference on e-Business Engineering (ICEBE) (2006)
14. Travers, J., Milgram: An experimental study of the small world problem. Sociometry 32(4), 425–443 (1969)
15. Turner, T.C., Smith, M.A., Fisher, D., Welser, H.T.: Picturing Usenet: Mapping Computer-Mediated Collective Action. Journal of Computer-Mediated Communication 10(4) (2005)

Segmentation and Automated Social Hierarchy Detection through Email Network Analysis*

Germán Creamer[1,3], Ryan Rowe[2], Shlomo Hershkop[3], and Salvatore J. Stolfo[3]

[1] Center for Computational Learning Systems, Columbia University,
New York, NY 10027
[2] Department of Applied Mathematics, Columbia University, New York, NY 10027
[3] Department of Computer Science, Columbia University, New York, NY 10027
{ggc14,rrr2107}@columbia.edu, {shlomo,sal}@cs.columbia.edu

Abstract. We present our work on automatically extracting social hierarchies from electronic communication data. Data mining based on user behavior can be leveraged to analyze and catalog patterns of communications between entities to rank relationships. The advantage is that the analysis can be done in an automatic fashion and can adopt itself to organizational changes over time.

We illustrate the algorithms over real world data using the Enron corporation's email archive. The results show great promise when compared to the corporations work chart and judicial proceeding analyzing the major players.

General Terms: Social Network, Enron, Behavior Profile, Link Mining, Data Mining, Corporate Householding.

1 Introduction

There is a vast quantity of untapped information in any collection of electronic communication records. Current techniques of manual sifting and hard coded keyword searches do not scale to the task of analyzing these collections. The recent bankruptcy scandals in publicly held US companies such as Enron and WorldCom, and the subsequent Sarbanes-Oxley Act have increased the need to analyze these vast stores of electronic information in order to define risk and identify any conflict of interest among the entities of a corporate household. Corporate household is 'a group of business units united or regarded united within the corporation, such as suppliers and customers whose relationships with the corporation must be captured, managed, and applied for various purposes' [23]. The problem can be broken into three distinct phases; entity identification, entity aggregation, and transparency of inter-entity relationships [22].

---

* This work is based on an earlier work: Automated Social Hierarchy Detection through Email Network Analysis in Proceedings of the 9th WebKDD and 1st SNA-KDD 2007 workshop on Web mining and social network analysis ACM, 2007. http://doi.acm.org/10.1145/1348549.1348562

H. Zhang et al. (Eds.): WebKDD/SNA-KDD 2007, LNCS 5439, pp. 40–58, 2009.

Identifying individual entities is straightforward process, but the relationships between entities, or corporate hierarchy is not a straightforward task. Corporate entity charts sometimes exist on paper, but they do not reflect the day to day reality of a large and dynamic corporation. Corporate insiders are aware of these private relationships, but can be hard to come by, especially after an investigation. This information can be automatically extracted by analyzing the email communication data from within a corporation.

Link mining is a set of techniques that uses different types of networks and their indicators to forecast or to model a linked domain. Link mining has been applied to many different areas [28] such as money laundering [17], telephone fraud detection [9], crime detection [31], and surveillance of the NASDAQ and other markets [17,13]. Perlich and Huang [26] show that customer modeling is a special case of link mining or relational learning [27] which is based on probabilistic relational models such as those presented by [12,34,35]. A recent survey of the literature can be found in [11]. In general models classify each entity independently according to its attributes. Probabilistic relational models classify entities taking into account the joint probability among them. The application of link mining to corporate communication is of course limited by restrictions to disseminate internal corporate data. Thus testing algorithms against real world data is hard to come by. An exception to this situation is the publicly available Enron email dataset.

The Enron Corporation's email collection described in section 2, is a publicly available set of private corporate data released during the judicial proceedings against the Enron corporation. Several researchers have explored it mostly from a Natural Language Processing (NLP) perspective [19,21,24]. Social network analysis (SNA) examining structural features [6] has also been applied to extract properties of the Enron network and attempts to detect the key players around the time of Enron's crisis; [7] studied the patterns of communication of Enron employees differentiated by their hierarchical level; [16] interestingly enough found that word use changed according to the functional position, while [5] conducted a thread analysis to find out employees' responsiveness. [30] used an entropy model to identify the most relevant people, [8] presents a method for identity resolution in the Enron email dataset, and [1] applied a cluster ranking algorithm based on the strength of the clusters to this dataset.

The work presented in this paper differs in two major ways. First, the relationship between any two users are calculated based on behavior patterns of each specific user not just links. This allows the algorithm to judge the strength of communication links between users based on their overall communication pattern. Second, we assume a corporate householding perspective and propose a methodology to solve the problem of transparency of inter-entity relationships in an automatic fashion. Our approach determines link mining metrics which can reproduce approximate social hierarchy within an organization or a corporate household, and rank its members. We use our metric to analyze email flows within an organization to extract social hierarchy. We analyze the behavior of the communication patterns without having to take into account the actual contents of the email messages.

By performing behavior analysis and determining the communication patterns we are able to automatically:

- Rank the major officers of an organization.
- Group similarly ranked and connected users in order to accurately reproduce the organizational structure in question.
- Understand relationship strengths between specific segments of users.

This work is a natural extension of previous work on the Email Mining Toolkit project (EMT) [32,33]. New functionality has been introduced into the EMT system for the purposes of automatically extracting social hierarchy information from any email collection.

The rest of the paper is organized as follows: Section 2 describes the Enron email corpus, section 3 presents the methods used to rank the Enron's officers; section 4 describes the research design; section 5 presents the results; section 5 discusses the results, and section 6 presents the conclusions.

2   Enron Antecedents and Data

The Enron email data set is a rich source of information showcasing the internal working of a real corporation over a period between 1998-2002. There seems to be multiple versions of the "official" Enron email data set in the literature [6,29,20,4]. In the midst of Enron's legal troubles in 2002, the Federal Energy Regulatory Commission (FERC) made a dataset of 619,449 emails from 158 Enron employees available to the public removing all attachment data. Cohen first put up the raw email files for researchers in 2004, the format was mbox style with each message in its own text file [4]. Following this, a number of research groups around the country obtained and manipulated the dataset in a variety of ways in attempts to correct inconsistencies and integrity issues within the dataset. Like [6], the version of the dataset we use to conduct our own research was treated and provided by Shetty and Adibi from ISI [29]. Our final main dataset has 149 users after cleaning. We call this dataset as the ENRON dataset. The ISI treatment of the Enron corpus consisted of deleting extraneous, unneeded emails and fixing some anomalies in the collection data having to do with empty or illegal user email names and bounced emails messages. In addition duplicates and blank emails were removed. We also used a suplementary file provided by [29] to assign the position of each user. When we apply the occupational classification suggested by the former authors to our dataset, we find that 38.5% of the users are classified as "employee" or "N/A". The classification "employee" does not bring any additional information more than indicating that the user is formally working at Enron. We reviewed the emails of those employees that were not well classified and imputed a position based on their signatures, the content of the email or lists of traders that circulated internally. We found out that an important part of the "unknown" employees were traders or were acting as traders.

We also used another segment of the major FERC dataset that includes only the emails among the 54 workers that we identified as members of the North American West Power Traders division. We called this dataset as TRADER. The importance of this dataset is that [25] presents an organigram of the above division.

It should be noted that [3] has found that there is indication that a significant number of emails were lost either in converting the Enron data set or through specific deletion of key emails. So although we are working with most of the emails, we will make the assumption that the algorithm is robust although some emails are not part of the analysis. In addition the FERC dataset only covers about 92% of Enron employees at the time.

3   SNA Algorithm

The social network analysis algorithm works as follows:

For each email user in the dataset analyze and calculate several statistics for each feature of each user. The individual features are normalized and used in a probabilistic framework with which users can be measured against one another for the purposes of ranking and grouping. It should be noted that the list of email users in the dataset represents a wide array of employee positions within the organization or across organizational departments.

Two sets of statistics are involved in making the decision about a given user's "importance." First, we collect information pertaining to the flow of information, both volumetric and temporal. Here we count the number of emails a user has sent and received in addition to calculating what we call the **average response time** for emails. This is, in essence, the time elapsed between a user sending an email and later receiving an email from that same user. An exchange of this nature is only considered a "response" if a received message succeeds a sent message within three business days. This restriction has been implemented to avoid inappropriately long response times caused by a user sending an email, never receiving a response, but then receiving an unrelated email from that same user after a long delay, say a week or two. These elapsed time calculations are then averaged across all "responses" received to make up the average response time.

Second, we gather information about the nature of the connections formed in the communication network. Here we rank the users by analyzing **cliques** (maximal complete subgraphs) and other graph theoretical qualities of an email network graph built from the dataset. Using all emails in the dataset, one can construct an undirected graph, where vertices represent accounts and edges represent communication between two accounts. We build such a graph in order to find all cliques, calculate degree and centrality measures and analyze the social structure of the network. When all the cliques in the graph have been found, we can determine which users are in more cliques, which users are in larger cliques, and which users are in more important cliques. We base it on the assumption that users associated with a larger set and frequency of cliques will then be ranked higher.

Finally all of the calculated statistics are normalized and combined, each with an individual contribution to an overall social score with which the users are ultimately ranked.

3.1 Information Flows

First and foremost, we consider the volume of information exchanged, i.e. the number of emails sent and received, to be at least a limited indicator of importance. It is fair to hypothesize that users who communicate more, should, on average, maintain more important placement in the social hierarchy of the organization. This statistic is computed by simply tallying the total number of emails sent and received by each user.

Furthermore, in order to rate the importance of user i using the amount of time user j takes to respond to emails from user i, we must first hypothesize that a faster response implies that user i is more important to user j. Additionally, when we iterate and average over all j, we will assume that the overall importance of user i will be reflected in this overall average of his or her importance to each of the other people in the organization. In other words, if people generally respond (relatively) quickly to a specific user, we can consider that user to be (relatively) important. To compute the average response time for each account x, we collect a list of all emails sent and received to and from accounts y_1 through y_n, organize and group the emails by account y_1 through y_n, and compute the amount of time elapsed between every email sent from account x to account y_j and the next email received by account x from account y_j. As previously mentioned, communication of this kind contributes to this value only if the next incoming email was received within three business days of the original outgoing email.

3.2 Communication Networks

The first step is to construct an undirected graph and find all cliques. To build this graph, an email threshold N is first decided on. Next, using all emails in the dataset, we create a vertex for each account. An undirected edge is then drawn between each pair of accounts which have exchanged at least N emails. We then employ a clique finding algorithm, Algorithm 457, first proposed by Bron and Kerbosch [2]. This recursively finds all maximal complete subgraphs (cliques).

a. *Number of cliques:* The number of cliques that the account is contained within.
b. *Raw clique score:* A score computed using the size of a given account's clique set. Bigger cliques are worth more than smaller ones, importance increases exponentially with size.
c. *Weighted clique score:* A score computed using the "importance" of the people in each clique. This preliminary "importance" is computed strictly from the number of emails and the average response time. Each account in a clique is given a weight proportional to its computed preliminary. The weighted clique score is then computed by adding each weighed user contribution within the clique. Here the 'importance' of the accounts in the clique raises the score of the clique.

More specifically, the raw clique score R is computed with the following formula:

$$R = 2^{n-1}$$

where n is the number of users in the clique. The weighted clique score W is computed with the following formula:

$$W = t \cdot 2^{n-1}$$

where t is the time score for the given user.

Finally, the following indicators are calculated for the graph $G(V, E)$ where $V = v_1, v_2, ..., v_n$ is the set of vertices, E is the set of edges, and e_{ij} is the edge between vertices v_i and v_j:

- Degree centrality or degree of a vertex v_i: $deg(v_i) \doteq \sum_j a_{ij}$ where a_{ij} is an element of the adjacent matrix A of G
- Clustering coefficient: $C \doteq \frac{1}{n} \sum_{i=1}^{n} CC_i$, where $CC_i \doteq \frac{2|\{e_{ij}\}|}{deg(v_i)(deg(v_i)-1)}$: $v_j \in N_i$, $e_{ij} \in E$. Each vertex v_i has a neighborhood N defined by its immediately connected neighbors: $N_i = \{v_j\} : e_{ij} \in E$.
- Mean of shortest path length from a specific vertex to all vertices in the graph G: $L \doteq \frac{1}{n} \sum_j d_{ij}$, where $d_{ij} \in D$, D is the geodesic distance matrix (matrix of all shortest path between every pair of vertices) of G, and n is the number of vertices in G.
- Betweenness centrality $B_c(v_i) \doteq \sum_i \sum_j \frac{g_{kij}}{g_{kj}}$. This is the proportion of all geodesic distances of all other vertices that include vertex v_i where g_{kij} is the number of geodesic paths between vertices k and j that include vertex i, and g_{kj} is the number of geodesic paths between k and j [10].
- "Hubs-and-authorities" importance: "hub" refers to the vertex v_i that points to many authorities, and "authority" is a vertex v_j that points to many hubs. We used the recursive algorithm proposed by [18] that calculates the "hubs-and-authorities" importance of each vertex of a graph $G(V, E)$.

3.3   The Social Score

We introduce the social score S, a normalized, scaled number between 0 and 100 which is computed for each user as a weighted combination of the number of emails, response score, average response time, clique scores, and the degree and centrality measures introduced above. The breakdown of social scores is then used to:

i.   Rank users from most important to least important
ii. Group users which have similar social scores and clique connectivity
iii. Determine n different levels (or echelons) of social hierarchy within which to place all the users. This is a clustering step, and n can be bounded.

The rankings, groups and echelons are used to reconstruct an organization chart as accurately as possible. To compute S , we must first scale and normalize each of the previous statistics which we have gathered. The contribution, C, of each metric is individually mapped to a [0, 100] scale and weighted with the following formula:

$$w_x \cdot C_x = w_x \cdot 100 \cdot \left[\frac{x_i - \inf x}{\sup x - \inf x} \right]$$

where x is the metric in question, w_x is the respective weight for that metric, the $\sup x$ and $\inf x$ are computed across all i users and x_i is the value for the user. This normalization is applied to each of the following metrics:

1. number of emails
2. average response time
3. response score
4. number of cliques
5. raw clique score
6. weighted clique score
7. degree centrality
8. clustering coefficient
9. mean of shortest path length from a specific vertex to all vertices in the graph
10. betweenness centrality
11. "Hubs-and-Authorities" importance

Finally, these weighted contributions are then normalized over the chosen weights w_x to compute the social score as follows:

$$S = \frac{\sum_{\text{all } x} w_x \cdot C_x}{\sum_{\text{all } x} w_x}$$

This gives us a score between 0 and 100 with which to rank every user into an overall ranked list. Our assumption is that although the number of emails, average response time, number and quality of cliques, and the degree and centrality measures are all perfectly reasonable variables in an equation for "importance," the appropriate contribution, i.e. weight, of each will vary by situation and organization, and therefore can be adjusted to achieve more accurate results in a variety of cases.

3.4   Visualization

As part of this research, we developed a graphical interface for EMT, using the JUNG library, to visualize the results of social hierarchy detection by means of email flow.

After the results have been computed, the statistics calculated and the users ranked, the option to view the network is available. When this option is invoked, a hierarchical, organized version of the undirected clique graph is displayed. Nodes represent users, while edges are drawn if those two users have exchanged at least m emails. Information is provided to the user in two distinct ways, the qualities of a user are reflected in the look of each node, where the relative importance of a user is reflected in the placement of each node within the simulated organization chart.

Although every node is colored red, its relative size represents its social score. The largest node representing the highest ranked individual, the smallest representing the lowest. The transparency of a given node is a reflection of the user's time score. A user boasting a time score near to 1 will render itself almost completely opaque where a user with a very low time score will render almost entirely transparent.

The users are divided into one of n echelons using a grouping algorithm, we use $n = 5$ in this paper. Currently, the only grouping algorithm which has been implemented is a straight scale level division. Users with social scores from 80-100 are placed on the top level, users with social scores from 60-80 are placed on the next level down, etc. If the weights are chosen with this scale division in mind, only a small percentage of the users will maintain high enough social scores to inhabit the upper levels, so a tree-like organizational structure will be manifested. Different, more sophisticated, ranking and grouping algorithms have been considered and will be implemented, and will be discussed in the following section on future work.

When a node is selected with the mouse, all users connected to the selected user through cliques are highlighted and the user, time score and social score populate a small table at the bottom of the interface for inspection. Nodes can be individually picked or picked as groups and rearranged at the user's discretion. If the organization is not accurate or has misrepresented the structure of the actual social hierarchy in question, the user can return to the analysis window and adjust the weights in order to emphasize importance in the correct individuals and then can recreate the visualization.

If the user would prefer to analyze the network graphically with a non-hierarchical structure, a more traditional graph/network visualization is available by means of the Fruchterman-Reingold node placement algorithm. This node placement algorithm will emphasize the clique structure and the connectedness of nodes in the graph rather than the hierarchical ranking scheme in the first visual layout.

4   Research Design

We ranked the employees of both datasets ENRON and TRADERS using the social score (see Table 5 and 6). We separated the ENRON dataset in four equal-sized segments where the top and low segments have the employees with the highest and lowest social scores respectively. We also classified the workers into four occupational categories:

1. Senior managers: CEO, chief risk officer, chief operating officer, presidents, vice presidents, and managing directors.
2. Middle managers: directors, managers, senior managers, lawyers, senior specialists, legal specialists, assistants to president, and risk management head. Assistants to president may qualify as regular "employees", however they communicate and take similar decisions to those that a middle manager may take.

Table 1. Actual distribution of Enron's employees in occupational categories and segments defined by the social score

	Sr.Mgrs	Mgrs	Traders	Employees	Total
1	20	11	0	6	37
2	11	8	10	8	37
3	5	14	11	7	37
4	3	12	15	8	38
Total	39	45	36	29	149

3. Traders. Some traders might be more important than a middle manager according to their performance, however we keep them in a separate category because of Enron's leadership as an energy trading company.
4. Employees: employees, employee associates, analysts, assistant traders, and administrative assistants.

We expect that there is a relationship between the occupational category and the segment that each employee belongs to. For instance, senior managers should be mostly in the first segment, and middle managers in the first and second segments. An exception is the last category because 23 workers still keep the generic title "employees." So they could be distributed among all the segments.

We built a 4 x 4 contingency table with the four segments and the four occupational categories (see Table 1). We wanted to test the hypothesis, using the Chi Square statistics, that there is a relationship between the occupational categories and the four segments of employees ranked by their social scores. So, we compared the ENRON contingency table with a contingency table that homogeneously distributes the same number of workers among the four segments (see Table 2). The null hypothesis is that the ENRON contingency table is not different from the expected contingency table.

In the case of the TRADERS dataset, the above analysis was not appropriate because it has fewer users and a flatter structure than the rest of the organization. We evaluated if the social score is capable of identifying the most important employees in the organizational structure or those that are in the top of the departamental organigram.

Table 2. Expected distribution of Enron's employees in occupational categories and segments defined by the social score

	Sr.Mgrs	Mgrs	Traders	Employees	Total
1	10	11	9	7	37
2	9	11	9	8	37
3	10	11	9	7	37
4	10	12	9	7	38
Total	39	45	36	29	149

5   Results and Discussion

We have performed the data processing and analysis using EMT [33]. EMT is a Java based email analysis engine built on a database back-end. The Java Universal Network/Graph Framework (JUNG) library [15] is used extensively in EMT for the degree and centrality measures, and for visualization purposes (see section 3.4).

In order to showcase the accuracy of our algorithm we present separate analysis of the complete Enron dataset and the North American West Power Traders division of Enron.

5.1   Analysis of Complete ENRON Dataset

In the case of the ENRON dataset, the Chi Square test rejects the null hypothesis with a probability of 99.6%. Hence, the four segments defined by the social score has also aggregated Enron's employees in a different way than a simple homogeneous distribution. In order to evaluate if the aggregation given by the social score also corresponds to the organizational hierarchy, we ranked the occupational groups in a scale of one to four based on a weighted average of the distribution of each occupational group in the four segments where one represents the highest hierarchy (see Table 3).

Table 3. Weighted ranking of each occupational category. The ranking is based on the distribution of each group of employees in the four hierarchical segments.

Occupational category	Weighted ranking
Senior managers	1.77
Middle managers	2.6
Traders	3.14
Employees	2.59

Table 3 shows a direct relationship of the ranking and the hierarchy of each occupational category, at exception of the generic category "employees" which has a ranking similar to the one of the middle managers. We suppose that this category may hide workers from other categories that were not well classified. Senior managers are present in the first (51.3%) and second (28.2%) segments of the ENRON contingency table (see Table 4), so their ranking is 1.77.

Middle managers have a ranking of 2.6. There is clearly a major jump with senior managers and their hierarchical level is higher than the one of the traders. The preeminence of Enron as an energy trading company leads to a slight distinction between the hierarchy of managers and traders. Even though managers organized the company, traders were the main drivers of the company. Therefore, the ranking of the traders is just slightly below the ranking of the managers.

Traders are mostly concentrated in the third and fourth segments (30.56% and 41.7% respectively) which is consistent with a ranking of 3.14. Most of the traders do not have a large number of emails. This can be explained because

Table 4. Proportional distribution of Enron's employees in occupational categories and segments defined by the social score

	Sr. Mgrs	Mgrs	Traders	Employees
1	51.28%	24.44%	0.00%	20.69%
2	28.21%	17.78%	27.78%	27.59%
3	12.82%	31.11%	30.56%	24.14%
4	7.69%	26.67%	41.67%	27.59%

Table 5. Social scores of employees. Enron North American subsidiary.

Name	Position	# Email	Avg Time	Response	# Cliques	RCS	WCS	Degree	Btw	Hubs	Avg.Dist.	CC	Score
Tim Beldon	Vice President	1266	2493	0.641	236	251140	1261588	83.00	370.35	0.04	1.00	0.40	75.68
Debora Davidson Sr.	Admin Assist	537	17	0.757	235	251136	1261586	66.00	278.35	0.04	1.02	0.41	63.51
Anna Meher	Admin Assist	544	1833	0.506	231	250368	1259149	62.00	260.94	0.04	1.04	0.42	62.84
Carla Hoffman	Staff	739	1319	0.576	221	249232	1255447	55.00	143.98	0.04	1.13	0.49	61.67
Cara Semperger	Specialist	693	2707	0.506	137	167232	859288	63.00	82.96	0.03	1.25	0.52	53.68
Diana Scholtes	Manager Cash	468	2443	0.496	124	203520	1061153	45.00	21.44	0.03	1.43	0.70	53.31
Sean Crandall	Director Cash	412	2151	0.478	91	126912	657157	42.00	40.04	0.03	1.42	0.62	43.64
Holden Salisbury	Specialist	400	951	0.723	83	104192	532137	49.00	37.29	0.03	1.40	0.61	43.03
Mark Fischer	Manager Cash	346	1580	0.553	75	125952	676349	34.00	15.56	0.02	1.49	0.72	42.90
Heather Dunton	Specialist	329	2530	0.442	60	88736	462950	43.00	51.56	0.03	1.40	0.59	39.51
Bill Williams III	Analyst	257	3254	0.326	49	81408	437255	36.00	25.12	0.03	1.47	0.68	37.98
Paul Choi	Manager	157	N/A	0	91	130112	624944	44.00	48.03	0.03	1.38	0.60	36.02
Tim Heizenrader	Director	268	843	0.645	50	56960	298395	33.00	19.45	0.02	1.55	0.71	35.56
Chris Foster	Director	210	1612	0.56	46	58624	283552	35.00	23.18	0.02	1.49	0.66	34.74
Donald Robinson	Specialist	214	1486	0.545	23	34688	203384	27.00	6.67	0.02	1.62	0.81	33.03
Jeff Richter	Manager Cash	208	4393	0.12	34	43456	200427	25.00	12.80	0.02	1.57	0.74	32.53
Mike Swerzbin	Vice Pres. Term	269	1752	0.517	23	36672	195602	31.00	14.80	0.02	1.57	0.70	32.51
Stewart Rosman	Director	118	1386	0.567	20	40448	206036	26.00	6.85	0.02	1.62	0.81	32.25
Julie Sarnowski	Staff	284	2289	0.428	43	43008	220023	28.00	25.94	0.02	1.53	0.63	32.14
Stacy Runswick	Staff	188	2837	0.356	25	24064	134823	32.00	11.12	0.02	1.58	0.74	31.83
Mike Purcell	Staff	139	1338	0.626	11	15360	91653	24.00	5.02	0.02	1.66	0.79	30.36
Chris Mallory	Analyst Cash	180	N/A	0	56	78720	383567	27.00	9.92	0.02	1.55	0.76	30.19
Tom Alonso	Manager Cash	302	N/A	0	42	67584	362249	26.00	9.89	0.02	1.55	0.75	29.67
Greg Wolfe	Vice President	116	N/A	0	59	81920	388975	35.00	25.82	0.02	1.47	0.65	29.23
Matt Motley	Manager Term	223	N/A	0	26	56320	292362	23.00	3.04	0.02	1.62	0.88	28.93
Kim Ward	Manager	147	3901	0.206	4	768	2437	13.00	0.39	0.01	1.81	0.95	28.92
Jesse Bryson	Specialist	71	2346	0.428	17	6720	29988	23.00	7.42	0.02	1.66	0.77	28.10
Phil Platter	Sr. Specialist Cash	205	N/A	0	54	66528	315399	33.00	34.34	0.02	1.49	0.63	27.90
John Forney	Manager	63	5194	0.007	33	13504	47359	29.00	24.06	0.02	1.53	0.61	27.69
Geir Solberg	Analyst	127	3157	0.299	19	5760	23945	23.00	7.59	0.02	1.66	0.73	27.67
Stanley Cocke	Specialist	79	2689	0.367	21	14976	62360	26.00	18.15	0.02	1.57	0.64	27.40
Ryan Slinger	Specialist	111	1151	0.597	9	1344	5467	18.00	3.79	0.01	1.74	0.78	27.10
John Mallowny	Manager	140	N/A	0	16	41728	224918	31.00	6.50	0.02	1.60	0.81	26.74
Kourtney Nelson	Analyst	167	N/A	0	41	36032	176304	29.00	21.81	0.02	1.53	0.63	23.97
Lisa Gang	Sr. Specialist Cash	120	N/A	0	12	13056	65253	22.00	7.37	0.02	1.64	0.75	21.34
Monika Causholli	Analyst	44	N/A	0	12	3072	10871	16.00	2.21	0.01	1.74	0.86	20.58
Kelly Axford	Sr. Receptionist	76	N/A	0	4	2560	13698	15.00	1.68	0.01	1.75	0.87	20.51
Holli Krebs	Director	39	N/A	0	2	256	966	9.00	0.08	0.01	1.85	0.96	20.33
Les Rawson	Sr. Specialist	79	N/A	0	16	6656	26614	23.00	7.65	0.02	1.66	0.74	20.19
Jeremy Morris	Analyst	66	N/A	0	6	1024	3597	12.00	0.87	0.01	1.79	0.89	20.09
Robert Anderson	Sr. Specialist	44	N/A	0	2	256	958	8.00	0.15	0.01	1.85	0.96	20.06
Smith Day	Sr. Specialist Cash	14	N/A	0	1	32	75	6.00	0.00	0.01	1.91	1.00	20.00
Mark Guzman	Specialist Term	159	N/A	0	14	5248	20018	18.00	6.84	0.01	1.68	0.75	19.97
Caroline Emmert	Specialist	45	N/A	0	3	1024	4138	12.00	0.84	0.01	1.79	0.91	19.90
Steve Swan	Manager	28	N/A	0	2	192	622	9.00	0.20	0.01	1.85	0.93	19.55
Maria Van Houten	Specialist	20	N/A	0	2	128	411	7.00	0.11	0.01	1.87	0.95	19.44
Cooper Richey	Associate	36	N/A	0	7	1536	5001	14.00	2.68	0.01	1.75	0.82	18.89

Note: CC: Clustering coefficient, Btw: Betweenness, Avg.Dist.: average distance, CC: clustering coefficient, Score: social score, Response: response score.

of the parallel communication systems of the traders (instantaneous message, phone, Bloomberg or trading terminal). They also communicate mostly among themselves, hence their social scores might be reduced in relation to the scores of the rest of the organization.

Employees are almost equally distributed in the last three segments and with smaller presence in the first segment. The even distribution of "employees" is easily explained by its generic category. According to the emails, many of them have

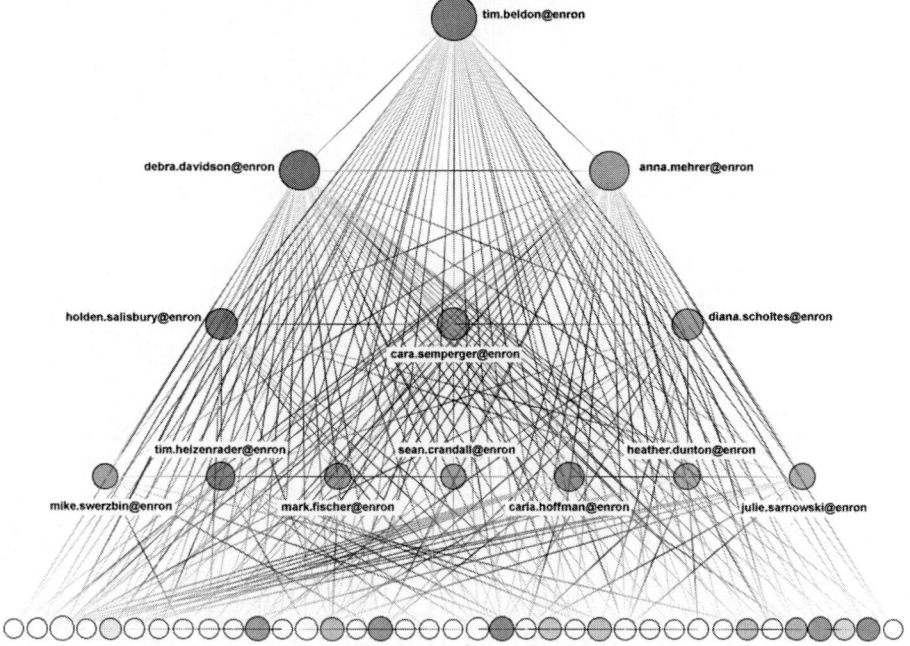

Fig. 1. Enron North American West Power Traders Extracted Social Network

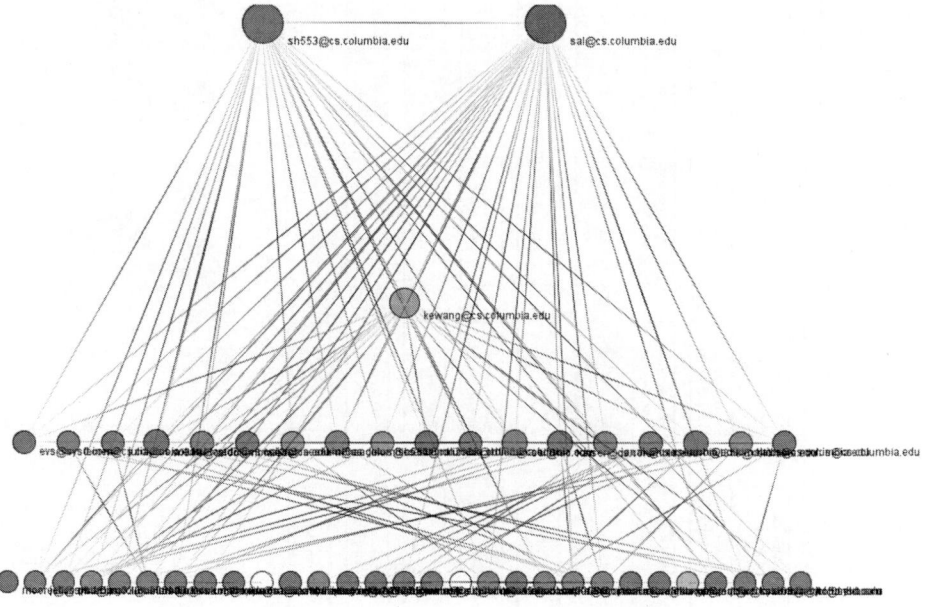

Fig. 2. Analysis of our own emails

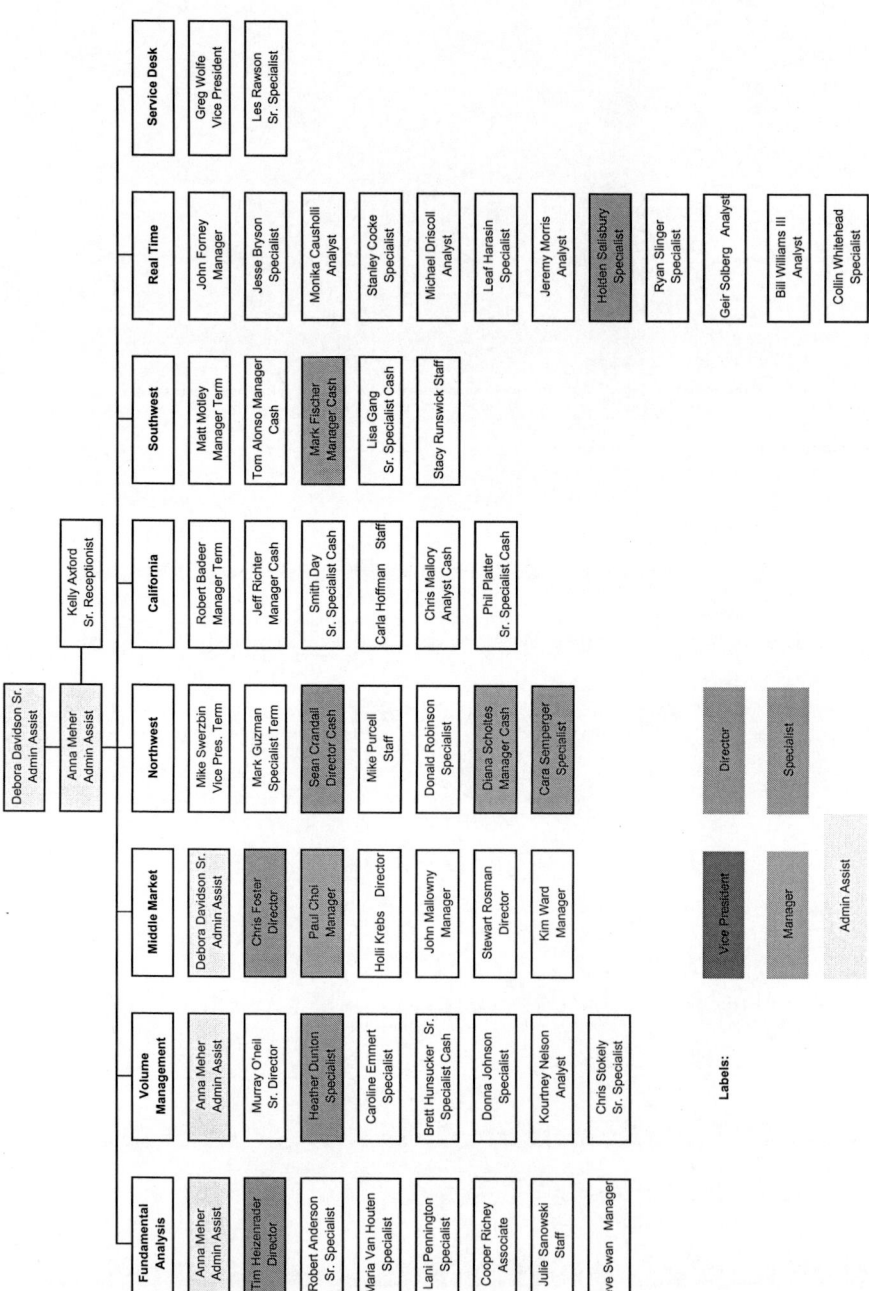

Fig. 3. Network Chart with highlighted results

Table 6. Social scores of employees. Enron.

Name	Position	# Email	Avg.Time	Response	# Cliques	RCS	WCS	Degree	Btw	Hubs	Avg.Dist.	CC	Score
Liz Taylor	Assistant to president	1724	1552	0.547	122	5790	24960	69	5087.36	0.03	1.57	0.12	64.76
Louise Kitchen	President, Enron Online	2480	1542	0.594	121	6452	27700	50	1459.04	0.03	1.76	0.21	60.07
Sally Beck	Chief operating officer	1107	2075	0.481	96	4978	22024	61	2899.41	0.03	1.70	0.14	54.63
Kenneth Lay	CEO, chairman	510	2986	0.347	98	5918	26441	52	1975.09	0.03	1.74	0.19	53.41
Michelle Lokay	Administrative assistant	1676	1583	0.568	16	6400	35576	16	85.37	0.00	2.52	0.75	47.31
Kimberly Watson	Employee	1803	1934	0.570	14	5890	33169	18	101.61	0.00	2.64	0.64	46.83
Lindy Donoho	Employee	997	1766	0.537	16	6400	35576	17	71.14	0.00	2.60	0.74	46.26
Steven Harris	Sr. manager *	1269	1515	0.565	16	6400	35576	18	104.80	0.00	2.50	0.69	46.20
Mike Grigsby	VP *	3463	1970	0.576	83	2096	6936	34	1006.34	0.02	1.89	0.28	44.11
Kevin Hyatt	Director, Pipeline business	623	1398	0.569	13	5248	29098	16	85.37	0.00	2.52	0.75	41.21
Lynn Blair	Manager *	975	1528	0.562	9	4480	25133	16	65.66	0.00	2.61	0.74	40.12
Drew Fossum	VP	949	1416	0.578	18	4892	27119	19	222.89	0.01	2.39	0.53	39.86
Teb Lokey	Regulatory affairs manager	324	1037	0.615	6	3200	18215	13	4.78	0.00	3.11	0.87	37.67
Jeff Dasovich	Director	4173	1737	0.765	22	856	3748	25	804.22	0.01	2.01	0.26	37.52
Darrell Schoolcraft	Manager *	371	2738	0.385	4	2816	16150	10	1.38	0.00	3.18	0.96	37.17
Shelley Corman	VP,regulatory affairs	922	1472	0.570	14	3624	19831	18	263.32	0.01	2.43	0.56	36.28
Scott Neal	VP -- trader	607	2353	0.437	53	2786	11015	31	616.46	0.02	1.97	0.29	35.89
Greg Whalley	President	613	611	0.670	32	3416	16964	21	122.24	0.02	2.02	0.47	35.63
Tana Jones	Senior legal specialist	3565	1374	0.683	25	644	2353	20	336.92	0.01	2.17	0.35	33.53
Rod Hayslett	VP, CFO and treasurer	853	1208	0.628	9	2816	15429	17	633.12	0.01	2.11	0.57	33.41
Andy Zipper	VP,Enron online	394	2389	0.430	41	2736	10828	16	49.78	0.01	2.15	0.56	33.26
Bill Rapp	Attorney	400	2104	0.478	5	2564	14670	12	101.66	0.00	2.70	0.79	33.26
Steven Kean	VP & chief of staff	740	1650	0.535	23	2744	14169	19	159.24	0.01	2.03	0.50	33.02
Sara Shackleton	VP	3343	1394	0.703	14	328	1192	12	34.45	0.00	2.56	0.59	32.89
Arnold John	VP	874	1233	0.597	41	2228	8488	20	165.76	0.01	2.07	0.43	32.76
Mark McConnell	Manager, TW *	692	1619	0.550	4	1536	8505	12	9.02	0.00	2.87	0.88	32.26
Rosalee Fleming	Employee	817	1840	0.506	11	2304	12269	17	41.10	0.01	2.05	0.62	31.88
Tracy Geaccone	Employee	614	771	0.676	5	1536	8481	12	11.48	0.00	2.82	0.83	31.83
Hunter Shively	VP -- trader	441	2013	0.483	33	2372	9822	20	212.59	0.01	2.09	0.41	31.70
Richard Shapiro	VP,regulatory affairs	2630	964	0.667	19	888	3853	18	182.02	0.01	2.14	0.45	31.46
Phillip Allen	Managing director	567	360	0.704	40	1960	7777	18	86.07	0.01	2.05	0.51	31.43
Paul Barbo	Sr. manager *	492	1973	0.489	4	2054	12093	13	192.58	0.00	2.68	0.68	31.04
Stanley Horton	President,Enron gas pipeline	410	1629	0.533	10	2576	14133	18	357.72	0.01	2.08	0.51	30.98
Jeffrey Skilling	Ceo	168	886	0.633	12	2536	13586	14	30.27	0.01	2.12	0.64	30.76
Rick Buy	Chief risk management	350	559	0.678	12	2072	11337	13	33.82	0.01	2.09	0.74	30.56
Mark Haedicke	Managing director,legal	414	1220	0.590	30	2000	10132	23	535.05	0.01	1.99	0.36	30.18
Stephanie Panus	Employee	1823	1999	0.551	30	608	2055	17	201.55	0.01	2.37	0.38	29.09
Matthew Lenhart	Trader	1112	1563	0.553	42	1060	3302	20	314.77	0.01	2.05	0.38	29.07
Marie Heard	Lawyer	1910	1735	0.574	19	552	2025	13	58.27	0.01	2.22	0.64	29.06
Susan Bailey	Legal specialist *	1623	2152	0.486	10	320	1157	10	7.11	0.00	2.67	0.76	29.03
Jeffrey Shankman	President,Enron global mkts	178	951	0.625	5	1680	9627	14	23.96	0.01	2.14	0.71	29.02
Kevin Presto	VP	659	1637	0.548	52	1016	2932	22	208.31	0.01	2.11	0.30	28.93
Kam Keiser	Employee	950	985	0.640	39	816	2428	24	352.67	0.01	2.09	0.30	28.56
Fletcher Sturm	VP	248	1364	0.568	26	1344	5640	17	53.18	0.01	2.11	0.52	28.27
Carol Clair	In house lawyer	1846	1194	0.683	11	192	627	9	36.69	0.00	2.53	0.64	28.20
Thomas Martin	VP	273	993	0.619	19	1736	7567	16	98.15	0.01	2.13	0.49	28.19
Susan Scott	Assistant trader *	1070	1891	0.507	28	780	2975	29	1277.64	0.01	2.05	0.14	27.85
Phillip Platter	Sr. specialist	51	222	0.721	2	32	89	4	0.00	0.00	3.09	1.00	26.82
Jay Reitmeyer	Trader *	579	2619	0.399	31	848	2607	15	175.84	0.01	2.14	0.48	26.68
Monique Sanchez	Trader *	663	1891	0.500	33	684	2020	24	307.16	0.01	2.08	0.27	26.52
James Steffes	VP,government affairs	1460	1161	0.666	17	384	1334	13	139.87	0.01	2.21	0.44	26.46
Barry Tycholiz	VP	1170	722	0.681	26	440	1317	16	146.66	0.01	2.07	0.43	26.32
Mike Maggi	Director	302	1031	0.612	9	648	2551	10	9.24	0.01	2.24	0.82	26.27
Phillip Love	Sr. manager -- traders *	588	1178	0.611	29	492	1335	23	504.59	0.01	2.19	0.26	26.12
Danny McCarty	VP	302	1138	0.600	4	1216	6720	14	201.46	0.01	2.18	0.58	25.87
Jane Tholt	VP	72	2127	0.463	2	12	24	3	0.00	0.00	2.86	1.00	25.83
Matt Smith	Trader *	390	2198	0.455	7	196	620	8	4.97	0.00	2.51	0.82	25.63
Elizabeth Sager	VP *	761	687	0.667	23	568	2073	21	389.14	0.01	2.09	0.34	25.61
Jason Williams	Trader *	792	1864	0.509	28	468	1418	17	220.80	0.01	2.10	0.39	25.24
Gerald Nemec	Attorney *	992	1746	0.547	26	364	1061	20	644.56	0.01	2.09	0.26	25.06
Mark Taylor	Employee	679	1180	0.610	7	368	1503	11	46.54	0.01	2.28	0.67	24.94
Debra Perlingiere	Legal specialist *	820	1666	0.549	22	422	1347	20	335.68	0.01	2.30	0.29	24.91
Cara Semperger	Senior analyst cash	334	1878	0.503	10	196	520	10	28.84	0.00	2.87	0.60	24.74
Jason Wolfe	Trader *	392	889	0.631	17	256	742	10	14.23	0.01	2.17	0.69	24.55
Tori Kuykendall	Trader	468	474	0.689	10	226	705	10	19.66	0.00	2.63	0.58	24.52
Errol McLaughlin	Employee	885	1687	0.527	10	228	710	13	52.74	0.01	2.24	0.51	24.31
John Griffith	Managing director UK	413	1522	0.547	7	400	1454	11	22.96	0.01	2.28	0.64	24.27
Lysa Akin	Employee	382	1696	0.526	29	412	1036	20	591.63	0.00	2.41	0.26	24.20
Stacy Dickson	Employee	510	1017	0.618	5	40	99	6	6.23	0.00	2.91	0.60	24.08
Theresa Staab	Employee	187	2365	0.435	2	24	59	5	3.22	0.00	2.74	0.80	23.99
Richard Sanders	VP,Enron wholesale svcs.	256	1596	0.538	11	204	654	9	17.24	0.00	2.60	0.56	23.49
Kate Symes	Trader *	292	1360	0.571	12	220	567	14	85.52	0.00	2.82	0.40	23.48
Kim Ward	Trader *	803	1082	0.629	22	222	596	16	464.85	0.01	2.02	0.25	23.43
Chris Germany	Trader *	972	2515	0.419	15	210	517	16	272.50	0.00	2.53	0.21	23.41
Dana Davis	VP term	245	2775	0.375	23	452	1151	11	41.85	0.01	2.16	0.47	23.40

Table 6. (*continued*)

Name	Position	# Email	Avg.Time	Response	# Cliques	RCS	WCS	Degree	Btw	Hubs	Avg.Dist.	CC	Score
Larry May	Director	303	2362	0.433	8	428	1628	14	121.11	0.01	2.11	0.52	23.40
Mark Whitt	VP *	698	1399	0.589	9	176	539	13	109.44	0.01	2.24	0.46	23.34
Frank Ermis	Director	338	548	0.678	14	252	772	12	27.19	0.01	2.18	0.56	23.26
Sandra Brawner	Director	160	614	0.671	8	352	1332	9	13.67	0.01	2.24	0.64	23.23
Keith Holst	Director	496	2829	0.368	16	424	1292	11	75.99	0.01	2.16	0.45	23.18
Randall Gay	Manager *	409	1052	0.610	12	266	820	11	170.52	0.01	2.16	0.56	23.14
Darron Giron	Trader *	205	1307	0.580	2	8	13	3	0.38	0.01	2.85	0.67	23.13
Mary Hain	In house lawyer	817	1014	0.632	20	224	568	18	452.53	0.01	2.21	0.23	23.01
Don Baughman	Director	277	2551	0.407	25	344	700	14	170.72	0.01	2.24	0.35	22.95
Jonathan Mckay	Director	207	640	0.666	20	504	1610	12	62.23	0.01	2.23	0.42	22.89
Michelle Cash	Legal specialist *	138	1891	0.497	8	224	760	9	17.51	0.00	2.48	0.58	22.85
Dan Hyvl	Employee	412	1495	0.562	7	92	279	9	57.62	0.00	2.59	0.53	22.83
Vince Kaminski	Risk management head	145	749	0.651	3	274	1380	7	34.56	0.01	2.35	0.67	22.71
Paul Lucci	Employee	254	1169	0.600	5	120	394	9	22.00	0.00	2.41	0.64	22.68
Doug Gilbert-smith	Manager	249	1500	0.551	14	192	470	12	44.13	0.01	2.22	0.47	22.62
Charles Weldon	Trader *	68	12	0.750	5	32	69	4	0.56	0.00	2.36	0.83	22.59
Harry Arora	VP	104	311	0.710	8	136	382	8	10.13	0.01	2.38	0.61	22.57
Bill Williams	Trader *	362	2271	0.446	10	176	473	14	266.01	0.01	2.72	0.33	22.54
Chris Dorland	Employee	202	669	0.662	28	372	990	16	115.63	0.01	2.11	0.28	22.12
Stacey White	Manager -- trader *	189	971	0.622	13	152	375	9	34.28	0.01	2.24	0.53	22.02
Vladi Pimenov	Trader *	85	1740	0.516	6	160	448	6	5.98	0.01	2.32	0.67	21.93
Eric Bass	Trader	632	962	0.641	15	172	470	14	204.18	0.01	2.09	0.32	21.93
Scott Hendrickson	Trader *	190	1535	0.544	3	24	37	4	6.28	0.00	2.52	0.67	21.84
Dutch Quigley	Trader *	440	1201	0.595	10	148	432	12	62.35	0.01	2.20	0.42	21.84
James Derrick	In house lawyer	236	1648	0.529	5	316	1406	10	75.22	0.01	2.14	0.53	21.83
Diana Scholtes	Trader	273	2185	0.458	10	176	495	11	250.41	0.01	2.30	0.51	21.78
Holden Salisbury	Cash analyst	132	1001	0.616	5	72	189	8	187.36	0.00	2.41	0.64	21.77
Kay Mann	Employee	374	1262	0.587	13	162	447	12	297.42	0.00	2.43	0.36	21.73
Martin Cuilla	Manager	215	1447	0.556	18	160	349	13	183.53	0.01	2.26	0.35	21.72
Geir Solberg	Trader *	111	1588	0.536	5	64	152	7	98.02	0.00	2.42	0.67	21.68
John Zufferli	VP *	186	835	0.639	12	112	247	8	18.01	0.01	2.26	0.54	21.57
John Hodge	Managing director	115	2081	0.470	3	36	76	5	18.37	0.01	2.61	0.60	21.54
Robin Rodrigue	Analyst *	50	1990	0.481	2	20	47	5	9.24	0.00	2.43	0.70	21.52
Judy Townsend	Trader *	323	1100	0.603	2	40	125	4	12.03	0.00	2.39	0.67	21.46
Sean Crandall	Director -- trader	214	1789	0.509	12	188	475	9	114.42	0.01	2.32	0.53	21.40
Mike Carson	Employee	112	597	0.671	9	112	282	6	7.63	0.01	2.22	0.60	21.20
Peter Keavey	Employee	54	11	0.750	2	18	46	4	1.12	0.00	2.78	0.50	21.15
Juan Hernandez	Senior specialist logistics	118	2266	0.445	5	40	59	6	8.68	0.00	2.39	0.60	21.12
Benjamin Rogers	Employee associate	82	304	0.711	4	40	82	5	5.95	0.00	2.45	0.60	21.05
Jim Schwieger	Trader	181	2022	0.478	14	348	1119	10	36.73	0.01	2.28	0.36	20.98
Joe Stepenovitch	VP,energy mkting & trading	112	5530	0.000	3	24	28	5	5.68	0.00	2.47	0.60	20.97
Eric Saibi	Trader	166	3086	0.332	5	52	113	6	11.82	0.00	2.26	0.60	20.81
Daren Farmer	Logistics manager	69	780	0.646	3	24	57	5	10.90	0.00	2.46	0.60	20.77
Robert Badeer	Director	404	500	0.684	6	38	84	7	59.38	0.00	2.49	0.43	20.63
Mike Swerzbin	Trader	156	1238	0.583	11	144	338	11	214.45	0.01	2.14	0.38	20.46
Ryan Slinger	Trader	122	1151	0.596	5	52	111	7	286.25	0.00	2.44	0.48	20.32
Eric Linder	Trader *	39	N/A	0.000	1	8	12	2	0.00	0.00	3.25	1.00	20.05
Kevin Ruscitti	Trader	74	968	0.621	3	16	29	4	13.60	0.00	2.89	0.33	19.93
John Forney	Manager,real time trading	211	3207	0.319	11	74	105	8	23.08	0.01	2.36	0.36	19.87
Lisa Gang	Director	73	N/A	0.000	1	16	35	4	0.00	0.00	3.12	1.00	19.77
Geoff Storey	Director -- trader	171	254	0.719	8	80	214	9	38.31	0.01	2.24	0.36	19.58
Richard Ring	Employee	41	973	0.620	2	10	22	3	5.51	0.00	2.85	0.33	19.45
Paul Thomas	Trader *	68	4685	0.115	6	60	92	8	76.00	0.00	2.32	0.39	19.30
Susan Pereira	Trader *	23	N/A	0.000	1	8	13	2	0.00	0.00	3.07	1.00	19.15
Joe Parks	Trader *	120	2490	0.414	3	28	62	5	28.88	0.00	2.42	0.40	18.87
Patrice Mims	Employee *	130	420	0.696	2	40	123	5	20.79	0.00	2.27	0.40	18.42
Steven South	Trader *	27	N/A	0.000	1	4	4	2	0.00	0.00	2.86	1.00	18.13
Chris Stokley	Employee *	68	3851	0.228	5	26	43	6	48.58	0.00	2.26	0.33	18.05
Mark Guzman	Trader	84	N/A	0.000	2	48	117	6	2.20	0.00	3.04	0.80	17.83
John Lavorato	CEO,Enron America	274	N/A	0.000	24	640	1846	16	71.23	0.01	2.24	0.28	16.08
Jeff King	Manager	94	N/A	0.000	11	96	146	5	1.04	0.00	2.34	0.80	15.63
David Delainey	CEO, Enron N.A. & E.energy	86	N/A	0.000	3	128	406	7	5.38	0.00	2.22	0.81	15.45
Tom Donohoe	Employee *	31	1287	0.577	2	6	7	2	4.08	0.00	2.61	0.00	15.11
Robert Benson	Director	100	N/A	0.000	7	112	218	6	3.08	0.01	2.29	0.73	14.95
Andrew Lewis	Director -- trader	74	N/A	0.000	3	32	60	5	1.83	0.01	2.32	0.80	14.82
Larry Campbell	Senior specialist	32	N/A	0.000	1	8	8	3	1.73	0.00	2.82	0.67	14.60
Albert Meyers	Trader *	37	814	0.641	2	4	4	2	13.73	0.00	2.50	0.00	14.54
Brad Mckay	Employee	120	N/A	0.000	10	176	379	9	35.19	0.01	2.34	0.56	14.40
Matt Motley	Director	90	N/A	0.000	6	84	179	6	35.49	0.00	2.58	0.53	13.50
Craig Dean	Trader	71	N/A	0.000	2	8	8	3	13.73	0.00	2.49	0.67	13.06
Andrea Ring	Trader *	118	N/A	0.000	3	22	37	6	88.75	0.00	2.32	0.47	11.56
Cooper Richey	Manager	91	N/A	0.000	5	36	55	7	46.05	0.00	2.26	0.43	11.24
Monika Causholli	Analyst risk management	30	N/A	0.000	2	6	4	2	7.91	0.00	2.57	0.00	6.51
Joe Quenet	Trader	37	N/A	0.000	3	10	6	2	0.77	0.00	2.50	0.00	6.31

Note: * indicates that the position is inferred by the content of emails. TW: Transwestern Pipeline Company, N.A.: North America, VP: vice president, E: Enron CC: Clustering coefficient, Btw: Betweenness, Avg.Dist.: average distance, CC: clustering coefficient, Score: social score, Response: response score.

a lot of influence in the company, however the emails studied did not indicate their occupational position. So we kept them in this generic category. When we eliminate this group of workers from our calculations, the probability of rejecting the null hypothesis using the Chi Square test is 99.95%.

5.2   Analysis of North American West Power Traders Division

As one can see in Table 5 and Figure 1, when running the code on the 54 users contained with the North American West Power Traders division we can reproduce the very top of the hierarchy with great accuracy. The transparency of the vertices in the graph visualization (Figure 1) denotes the response score of the user, a combination of the number of responses and the average response time. By our assumptions made in section three, we have determined that lower average response times infer higher importance, and appropriately, Tim Belden and Debra Davidson have fast average response times, causing more opaque colored node representations.

Once we turn to the lower ranked individuals, differences in our computed hierarchy and the official hierarchy are quite noticeable in Figure 3. As we move down the corporate ladder, the conversational flows of dissimilar employees can in fact be quite similar. Despite the discrepancies of our selections with the lower ranked officers, we find that consistently we are able to pick out the most important 2 or 3 individuals in any given segment, affording us the power to build a hierarchy from small groups up. Not only does the head of Enrons Western trading operation, Tim Belden, appear on the top of our list, both his administrative assistants appear with him. Additionally, in the first fourteen positions we are also able to identify the majority of directors, and an important number of managers and specialists. Figure 3 highlights these positions and their key role in the organizational structure.[1]

The placement of accounts other than the top two or three is in fact giving us insight into the true social hierarchy of this particular Enron business unit over the course of time from which the emails were gathered. This differs noticeably from the official corporate hierarchy, which can be expected as the data reflects the reality of the corporate communication structure.

With this sort of technique, it may be possible to view a snapshot of a corporate household or community (or any number of sub-communities) and effectively determine the real relationships and connections between individuals, a set of insights an official corporate organization chart simply could not offer.

6   Conclusions and Future Work

Understandingly, real world organizational data is hard to come by because of privacy concerns. The data in the the Enron dataset provides an excellent starting point for testing tools in a general setting. When we analyzed the algorithm

---

[1] Researchers interested in this line of research can find organigrams of public companies in their annual reports.

on our own email data the social hierarchy of our lab was very apparent. Figure 2 clearly shows professor, PhD, lab students, and outsiders.

In our analysis of the Enron dataset, we have been able to recognize and rank the major officers, group them by their hierarchy, and capture the relationship among the segment of users. We think that this approach contributes to the definition of corporate household in the case of Enron, and can be easily extended to other corporations.

The next immediate concern is to apply these tools to the Enron dataset in a comprehensive and formal manner over time based data sets. The dataset contains enough email volume and generality to provide us with very useful results if we are interested in knowing how social structure changes over time. By varying the feature weights it is possible to use the mentioned parameters to:

a. Pick out the most important individual(s) in an organization,
b. Group individuals with similar social/email qualities, and
c. Graphically draw an organization chart which approximately simulates the real social hierarchy in question

In order to more completely answer our question, as previously mentioned, a number of additions and alterations to the current algorithms exist and can be tested. First, the concept of average response time can be reworked or augmented by considering the order of responses, rather than the time between responses, like in [14]. For example, if user a receives an email from user b before receiving an email from user c, but then promptly responds to user c before responding to user b, it should be clear that user c carries more importance (at least in the eyes of user a). Either replacing the average response time statistic with this, or introducing it as its own metric may prove quite useful.

Another approach is to consider common email usage times for each user and to adjust the received time of email to the beginning of the next common email usage time. For example, if user a typically only accesses her email from 9-11am and from 2-5pm, then an email received by user a at 7pm can be assumed to have been received at 9am the next morning. We hypothesize that this might correct errors currently introduced in the average response time calculations due to different people maintaining different work schedules.

In addition to the continued work on the average response time algorithms, new grouping and division algorithms are being considered. Rather than implementing the straight scale division algorithm, a more statistically sophisticated formula can be used to group users by percentile or standard deviations of common distributions. Furthermore, rather than ignoring the clique connections between users at this step, the graph edges could very well prove important in how to arrange users into five different levels of social ranking, by grouping users with respect to their connections to others.

References

1. Bar-Yossef, Z., Guy, I., Lempel, R., Maarek, Y.S., Soroka, V.: Cluster ranking with an application to mining mailbox networks. In: ICDM 2006, pp. 63–74. IEEE Computer Society, Washington (2006)
2. Bron, C., Kerbosch, J.: Algorithm 457: finding all cliques of an undirected graph. Commun. ACM 16(9), 575–577 (1973)
3. Carenini, G., Ng, R.T., Zhou, X.: Scalable discovery of hidden emails from large folders. In: KDD 2005: Proceeding of the eleventh ACM SIGKDD international conference on Knowledge discovery in data mining, pp. 544–549. ACM Press, New York (2005)
4. Cohen, W.: Enron data set (March 2004)
5. Garg, D., Deepak, P., Varshney, V.: Analysis of Enron email threads and quantification of employee responsiveness. In: Proceedings of the Text Mining and Link Analysis Workshop on International Joint Conference on Artificial Intelligence, Hyderabad, India (2007)
6. Diesner, J., Carley, K.: Exploration of communication networks from the Enron email corpus. In: Proceedings of Workshop on Link Analysis, Counterterrorism and Security, Newport Beach, CA (2005)
7. Diesner, J., Frantz, T.L., Carley, K.M.: Communication networks from the Enron email corpus. Journal of Computational and Mathematical Organization Theory 11, 201–228 (2005)
8. Elsayed, T., Oard, D.W.: Modeling identity in archival collections of email: a preliminary study. In: Third Conference on Email and Anti-spam (CEAS), Mountain View, CA (July 2006)
9. Fawcett, T., Provost, F.: Activity monitoring: noticing interesting changes in behavior. In: Proceedings of the Fifth ACM SIGKDD International conference on knowledge discovery and data mining (KDD 1999), pp. 53–62 (1999)
10. Freeman, L.: Centrality in networks: I. conceptual clarification. Social networks 1, 215–239 (1979)
11. Getoor, L., Diehl, C.P.: Link mining: A survey. SIGKDD Explorations 7(2), 3–12 (2005)
12. Getoor, L., Friedman, N., Koller, D., Taskar, B.: Learning probabilistic models of link structure. Journal of Machine Learning Research 3, 679–707 (2002)
13. Goldberg, H.G., Kirkland, J.D., Lee, D., Shyr, P., Thakker, D.: The NASD securities observation, news analysis and regulation system (sonar). In: IAAI 2003 (2003)
14. Hershkop, S.: Behavior-based Email Analysis with Application to Spam Detection. PhD thesis, Columbia University (2006)
15. Joshua O'Madadhain, D.F., White, S.: Java universal network/graph framework. JUNG 1.7.4 (2006)
16. Keila, P., Sillicorn, D.: Structure in the Enron email dataset. Journal of Computational and Mathematical Organization Theory 11, 183–199 (2005)
17. Kirkland, J.D., Senator, T.E., Hayden, J.J., Dybala, T., Goldberg, H.G., Shyr, P.: The NASD regulation advanced detection system (ads). AI Magazine 20(1), 55–67 (1999)
18. Kleinberg, J.: Authoritative sources in a hyperlinked environment. Journal of the ACM 46 (1999)
19. Yeh, J.-Y., Harnly, A.: Email Thread Reassembly Using Similarity Matching. In: CEAS 2006 - Third Conference on Email and Anti-spam, July 27-28, 2006, Mountain View, California, USA (2004)

20. Klimt, B., Yang, Y.: Introducing the Enron corpus. In: CEAS (2004)
21. Klimt, B., Yang, Y.: The Enron corpus: A new dataset for email classification re-search. In: Boulicaut, J.-F., Esposito, F., Giannotti, F., Pedreschi, D. (eds.) ECML 2004. LNCS, vol. 3201, pp. 217–226. Springer, Heidelberg (2004)
22. Madnick, S., Wang, R., Xian, X.: The design and implementation of a corporate householding knowledge processor to improve data quality. Journal of Management Information Systems 20(3), 41–69 (Winter 2003)
23. Madnick, S., Wang, R., Zhang, W.: A framework for corporate householding. In: Fisher, C., Davidson, B. (eds.) Proceedings of the Seventh International Conference on Information Quality, Cambridge, MA, pp. 36–40 (November 2002)
24. McCallum, A., Corrada-Emmanuel, A., Wang, X.: The author-recipient-topic model for topic and role discovery in social networks: Experiments with Enron and academic email. In: NIPS 2004 Workshop on Structured Data and Representations in Probabilistic Models for Categorization, Whistler, B.C. (2004)
25. McCullough, R.: Memorandum related to reading Enron's scheme accounting materials (2004), http://www.mresearch.com/pdfs/89.pdf
26. Perlich, C., Huang, Z.: Relational learning for customer relationship management. In: Proceedings of International Workshop on Customer Relationship Management: Data Mining Meets Marketing (2005)
27. Perlich, C., Provost, F.: Acora: Distribution-based aggregation for relational learning from identifier attributes. Journal of Machine Learning (2005)
28. Senator, T.E.: Link mining applications: Progress and challenges. SIGKDD Explorations 7(2), 76–83 (2005)
29. Shetty, J., Adibi, J.: The Enron email dataset database schema and brief statistical report (2004)
30. Shetty, J., Adibi, J.: Discovering important nodes through graph entropy: the case of Enron email database. In: ACM SIGKDD International Conference on Knowledge Discovery and Data Mining, Chicago, Ill (August 2005)
31. Sparrow, M.: The application of network analysis to criminal intelligence: an assessment of the prospects. Social networks 13, 251–274 (1991)
32. Stolfo, S., Creamer, G., Hershkop, S.: A temporal based forensic discovery of electronic communication. In: Proceedings of the National Conference on Digital Government Research, San Diego, California (2006)
33. Stolfo, S.J., Hershkop, S., Hu, C.-W., Li, W.-J., Nimeskern, O., Wang, K.: Behavior-based modeling and its application to email analysis. ACM Transactions on Internet Technology 6(2), 187–221 (2006)
34. Taskar, B., Segal, E., Koller, D.: Probabilistic classification and clustering in relational data. In: Nebel, B. (ed.) Proceeding of IJCAI 2001, 17th International Joint Conference on Artificial Intelligence, Seattle, US, pp. 870–878 (2001)
35. Taskar, B., Wong, M., Abbeel, P., Koller, D.: Link prediction in relational data. In: Proceedings of Neural Information Processing Systems, 2004 (2004)

Mining Research Communities in Bibliographical Data*

Osmar R. Zaïane, Jiyang Chen, and Randy Goebel

University of Alberta, Canada
{zaiane,jiyang,goebel}@cs.ualberta.ca

Abstract. Extracting information from very large collections of structured, semi-structured or even unstructured data can be a considerable challenge when much of the hidden information is implicit within relationships among entities in the data. Social networks are such data collections in which relationships play a vital role in the knowledge these networks can convey. A bibliographic database is an essential tool for the research community, yet finding and making use of relationships comprised within such a social network is difficult. In this paper we introduce **DBconnect**, a prototype that exploits the social network coded within the **DBLP** database by drawing on a new random walk approach to reveal interesting knowledge about the research community and even recommend collaborations.

1 Introduction

A social network is a structure made up of nodes, representing entities from different groups, that are linked with different types of relations. Viewing and understanding social relationships between individuals or other entities is known as *Social Network Analysis* (SNA). SNA methods [26] are used to study organizational relations, analyze citation or computer mediated communications, etc. There are many applications such as studying the spread of disease, understanding the flow of communication within and between organizations, and so on. As an important field in SNA, *Community Mining* [5,16] has received considerable attention over the last few years. A community can be defined as a group of entities that share similar properties or connect to each other via certain relations. Identifying these connections and locating entities in different communities is an important goal of community mining and can also have various applications. We are interested in the application for finding potential collaborators for researchers by discovering communities in an author-conference social network, or recommending books (or other products) for users based on the borrowing records of other members of their communities in a library system. In this paper we are focusing on the social network implicit in the DBLP database which includes information about authors, their papers and the conferences they published in. DBLP [13,3] is an on-line resource providing bibliographic information on major computer science conference proceedings

---

* This work is based on an earlier work: DBconnect: mining research community on DBLP data, in Proceedings of the 9th WebKDD and 1st SNA-KDD 2007 workshop on Web mining and social network analysis, COPYRIGHT ACM, 2007, http://portal.acm.org/citation.cfm?doid=1348549.1348558

H. Zhang et al. (Eds.): WebKDD/SNA-KDD 2007, LNCS 5439, pp. 59–76, 2009.
© Springer-Verlag Berlin Heidelberg 2009

and journals[1]. It is such an essential index for the community that it was included in the ACM SIGMOD Anthology[2].

In SNA, the closeness of two related concepts in the network is usually measured by a relevance score, which is based on selected relationships between entities. It can be computed with various techniques, e.g., *Euclidean distance* or *Pearson correlation* [26]. Here we use the random walk approach to determine the relevance score between two entities. A random walk is a sequence of nodes in a graph such that when moving from one node n to the subsequent one in the sequence, one of n's neighbours is selected at random but with an edge weight taken into account. The closeness of a node b with respect to a node a is the static steady-state probability that the sequence of the nodes would include b when the random walk starts in a. This probability is computed iteratively until convergence, and is used as an estimated relevance score. In this paper, we adapt a variation of this idea, which is the random walk with restart (RWR): given a graph and a starting node a, at each step, we move to a neighbour of the current node at random, proportionally to the available edge weights, or go back to the initial node a with a restart probability c. RWR has been applied to many fields, e.g. anomaly detection [23], automatic image captioning [18], etc.

In this paper, we use DBLP data to generate bipartite (author-conference) and tripartite (author-conference-topics) graph models, where topics are frequent n-grams extracted from paper titles and abstracts. Moreover, we present an iterative random walk algorithm on these models to compute the relevance score between authors to discover the communities. We take into consideration the co-authorship while designing graphical models and the algorithm. We also present our ongoing work DBconnect, which provides an interactive interface for navigating the DBLP community structure online, as well as recommendations and explanations for these recommendations.

The rest of the paper is organized as follows. We discuss related work in Section 2. Graph models and Random walk algorithms for computing the relevance score are described in Section 3. The result and the ongoing DBconnect work is reported in Section 4 before the paper is concluded in Section 5.

2   Related Work

Community Mining

The ability to find communities within large social networks could be of important use, e.g., communities in a biochemical network might correspond to functional units of the same kind [8]. Since social networks can be easily modeled as graphs, finding communities in graphs, where groups of vertices within which connections are dense, but between which connections are sparser, has recently received considerable interests. Traditional algorithms, such as the spectral bisection method [19], which is based on the eigenvectors of the graph Laplacian, and the Kernighan-Lin algorithm [11], which greedily optimizes the number of within- and between-community edges, suffer from the fact that they only bisect graphs. While a larger number of communities

---

[1] In December 2007, DBLP comprised more than 970,000 entries.
[2] http://acm.org/sigmod/dblp/db/anthology.html

can be identified by repeated bisection, there is no hint of when to stop the repeated partitioning process. Another approach to find communities is hierarchical clustering based on similarity measures between objects, but it cannot handle the case where some vertices are not close enough to any of the principal communities to be clustered. In the last few years, several methods have been developed based on iterative removal of between-community edges [5,20,25]. Important results on researcher community mining have been revealed by analysis of a co-relation (e.g., co-authorship in a paper or co-starring in a movie) graph. Nascimento et al. [15] and Smeaton et al. [21] show co-authorship graphs for several selected conferences are small world graphs[3] and calculate the average distance between pairs of authors. Similarly, the Erdös Number Project[4] and the Oracle of Bacon[5] compute the minimum path length between one fixed person and all other people in the graph.

Community Information System
A related contribution in the context of recommending future collaborators based on their communities is W-RECMAS, which is an academic recommendation system developed by Cazella et al. [14]. The approach is based on collaborative filtering on the user profile data of the Brazilian e-government's database and can aid scientists by identifying people in the same research field or with similar interests in order to help exchange ideas and create academic communities. However, researchers need to post and update their research interests and personal information in the database before they can be recognized and recommended by the system, which makes the approach impractical. In order to efficiently browse the DBLP bibliographical database [13], Klink et al. [12] developed a specialized tool, DBL-Browser, which provides an interactive user interface and essential functionalities such as searching and filtering to help the user navigate through the complex data of DBLP. Another project to explore information for research communities is the DBLife system[6]. It extracts information from web resources, e.g., mailing list archives, newsletters, well-known community websites or research homepages, and provides various services to exploit the generated entity relationship graph [4]. While they do not disclose the process and the means used, they provide related researchers and related topics to a given researcher. In addition to the DBLife project supported by Yahoo, Microsoft Research Asia also developed a similar project called Libra[7], which discovers connected authors, conferences and journals etc. However, in our own experience of using the two systems, we found some incorrect instances of these related entities. Distinct from DBLife and Libra, our DB-connect focuses on finding related researchers more accurately based on a historical publication database and explicit existing relationships in the DBLP coded social network. Moreover, DBlife and Libra do not provide recommendations like DBconnect does.

---

[3] A small-world graph is a graph in which most nodes are not neighbors of one another, but can be reached from every other by a small number of hops or steps [2].
[4] http://www.oakland.edu/~grossman/erdoshp.html
[5] http://www.oracleofbacon.org/
[6] http://dblife.cs.wisc.edu/
[7] http://libra.msra.cn/

Random Walk Algorithm

As a popular metric to measure the similarity between entities, the random walk algorithm has received increasing attention after the undeniable success of the Google search engine, which applies a random walk approach to rank web pages in its search result as well as the list of visited pages to re-index [1]. Specifically, Page-Rank [17] learns ranks of web pages, which are N-dimensional vectors, by using an iterated method on the adjacency matrix of the entire web graph. In order to yield more accurate search results, Topic-Sensitive PageRank [6] pre-computes a set of biased PageRank vectors, which emphasize the effect of particular representative topic keywords to increase the importance of certain web pages. Those are used to generate query-specific importance scores. Alternatively, SimRank [9] computes a purely structural score that is independent of domain-specific information. The SimRank score is a structure similarity measure between pairs of pages in the web graph with the intuition that two pages are similar if they are related by similar pages. Unfortunately, SimRank is very expensive in computation since it needs to calculate similarities between many pairs of objects. According to the authors, a pruning technique is possible to approximate SimRank by only computing a small part of the object pairs. However, it is very hard to identify the right pairs to compute at the beginning, because the similarity between objects may only be recognized after the score between them is calculated. Similar random walk approaches have been used in other domains. For example, the Mixed Media Graph [18] applies a random walk on multimedia collection to assign keywords to the multimedia object, such as images and video clips, but a similarity function for each type of the involved media is required from domain experts. He et al. [7] propose a framework named MRBIR using a random walk on a weighted graph for images to rank related images given an image query. Sun et al. [23] detect anomaly data for datasets that can be modeled as bipartite graphs using the random walk with restart algorithm. Recent work by Tong et al. [24] proposed a fast solution for applying random walk with restart on large graphs, to save pre-computation cost and reduce query time with some cost on accuracy. While random walk algorithms such as SimRank computes links recursively on all pairs of objects, LinkClus [28] takes advantage of the power law distribution of links, and develops a hierarchical structure called SimTree to represent similarities in a multi-granularity manner. By merging computations that go through the same branches in the SimTree, LinkClus is able to avoid the high cost of pairwise similarity computations but still thoroughly explores relationships among objects without random walk.

In this paper, we apply a random walk approach on tripartite graphs to include topic information, and increase the versatility of the random walk by expanding the original graph model with virtual nodes that take the co-authorship into consideration for the DBLP data. These extensions are explained in the following section.

3   Proposed Method

Searching for relevant conferences, similar authors, and interesting topics is more important than ever before, and is considered an essential tool by many in the research community such as finding reviewers for journal papers or inviting program committee members for conferences. However, finding relationships between authors and thematically similar publications is becoming more difficult because of the mass of information

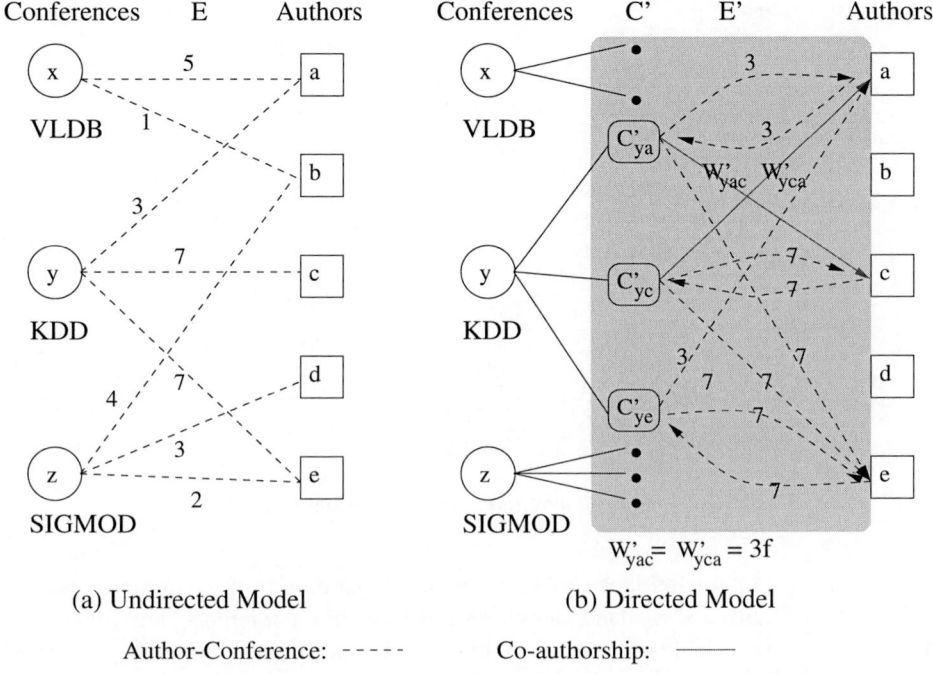

(a) Undirected Model   (b) Directed Model

Author-Conference: - - - - -   Co-authorship: ———

Fig. 1. Bipartite Model for Conference-Author Social Network

and the rapid growth of the number of scientific workers [12]. Moreover, except direct co-authorships which are explicit in the bibliographical data, relationships between nodes in this complex social network are difficult to detect by traditional methods. In order to understand relations between entities and find accurate researcher communities, we need to take into consideration not only the information of who people work with, i.e. co-authors, but also where they submit their work to, i.e., conferences, and what they work on, i.e. topics. In this section, we first present the models that incorporate these concepts, then discuss the algorithms that compute the relevance scores for these models.

Given the DBLP database $D = (C \cup A)$, where conference set $C = \{c_i | 1 \leq i \leq n\}$ and author set $A = \{a_j | 1 \leq j \leq m\}$, we can model D as an undirected bipartite graph $G = (C, A, E)$: conference nodes and author nodes are connected if the corresponding author published in the conference and there are no edges in E within the same group of nodes, i.e., author to author or conference to conference. Figure 1 (a) shows an example of the bipartite graph, representing social relationships between conference and author entities. The weights of the edges are publishing frequency of different authors in a certain conference.

3.1 Adding Topic Information

As mentioned before, the research topic is an important component to differentiate any research community. Authors that attend the same conferences might work on various

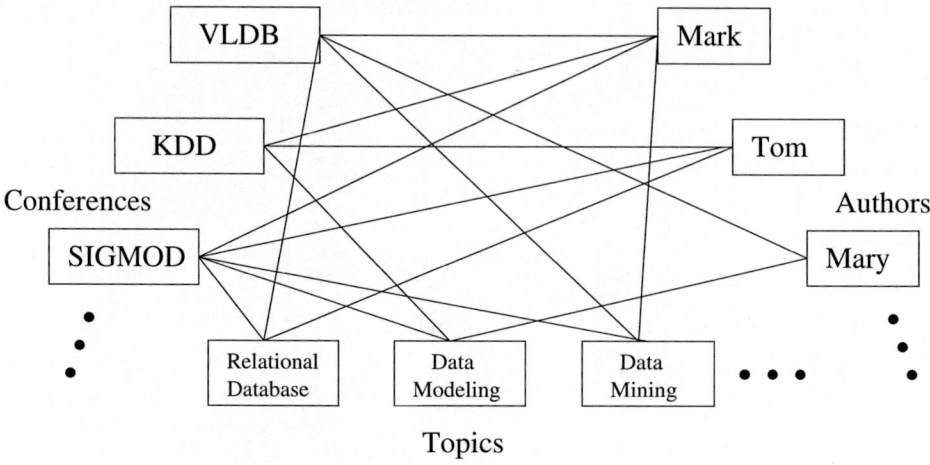

Fig. 2. Tripartite Graph Model for Author-Conference-Topic

topics. Therefore, topic entities should be treated separately from conference and author entities. Figure 2 shows an example of linked author, conference, and topic entities. DBLP contains table of contents of some conference proceedings. These table of contents include session titles that could be considered as topics. Unfortunately, very few conference proceedings have their table of contents included in DBLP, and in the affirmative, session titles are often absent. To extract relevant topics from DBLP we resorted to the paper titles instead. Moreover, we obtained as many paper abstracts as possible from Citeseer[8], then extracted topics based on keyword frequency from both titles and abstracts. We found that frequent co-locations in title and abstract text constitute reliable representation of topics. We concede that other methods are possible to get effective research topics.

We now consider a publication database $D = (C \cup A \cup T)$, where topic set $T = \{t_i | 1 \leq i \leq l\}$. We naturally use a tripartite graph to model such data: author/conference nodes are related to a topic node if they have a paper on that topic, the edge weight is the topic matching frequency. We apply the random walk algorithm on a tripartite graph by adjusting the walking sequence. For example, previously the random walker turns back to author nodes when it reaches a conference node; now it will go forward to topic nodes first, and then walk back to author nodes. By such modifications, the relevance score now contains both conference and topic influences, i.e., in a tripartite model, authors with high relevance score share similar conference experiences and paper topics with the given author.

3.2   Adding Co-author Relations

Table 1 shows the number of publications of five authors a, b, c, d, e in three conferences VLDB KDD and SIGMOD. Authors a and c have co-authored 3 papers in KDD,

---

[8] http://citeseer.ist.psu.edu/

Table 1. Author Publication Records in Conferences. For example, a, b, c, d, e are authors, $ac(3)$ means that author a and c published three papers together in a certain conference.

	Publication Records
VLDB(x)	a(4), ab(1)
KDD(y)	ac(3), c(4), e(7)
SIGMOD(z)	b(4), d(1), de(2)

a and b co-authored 1 paper in VLDB and d, e co-authored 2 papers in SIGMOD. Unfortunately, the corresponding bipartite graph, which is shown in Figure 1 (a), fails to represent any co-authorships. For example, author a and c co-authored many papers at KDD, but there are no edges in the bipartite graph that can be used to represent this information: edge $e(y, a)$ and $e(y, c)$ are both used by relations between conference and author. On the other hand, author e seems more related to author c since the weights of edges connecting them to KDD are the heaviest ($w_{yc} = 7$, $w_{ye} = 7$). The influence of the important co-author a is neglected because the model only represents publication frequency.

To capture the co-author relations, just adding a link between a and c does not suffice, since it misses the role of KDD, where the co-authorship happens. Making the link connecting a and c to KDD directional does not work either, as from KDD there are edges to many other authors, which would make the random walk infeasible (i.e., yielding undesirable results). Moreover, adding additional nodes to represent each co-author relation is impractical when there is a huge number of such relations. For instance, adding "Papers" between Authors and Conferences to make a tri-partite graph would actually not only add a significant number of edges since many authors have multiple papers per conference series, but also, this scheme does not allow the random walk to favor co-authorship as any author or co-author gets the same probability to be visited.

Our approach is to re-structure the bipartite model by adding surrogate nodes to replace the KDD node and having them link to a and c so that the random walk calculation can be applied while the connection between related nodes remains the same. In more detail, we add a virtual level of nodes to replace the conference partition, and add direction to the edges. Figure 1 (b) shows details of node KDD as an example. We first split y into 3 nodes to represent relations between y and authors who published there (a, c and e). These author nodes then connect to their own splitted relation nodes with the original weight ($e'(a, C'_{ya}), e'(c, C'_{yc}), e'(e, C'_{ye})$). Then we connect from C' nodes to all author nodes that have published at KDD. If the author node has a co-author relation with the author included in the C' node, the edge is weighted by co-author frequency multiplied by a parameter f (which is explained in the following), otherwise, the edge is weighted as original. We can see that the co-authorship, which is missed in the simple bipartite graph, is now represented by extra weight of edge $e'(C'_{yc}, a)$ and $e'(C'_{ya}, c)$, which shows author a is more related to c then author e through KDD due to their collaborations. The parameter f is used to control the co-author influence, usually we set $f = k$ (k is the total author number of a conference).

3.3   Random Walk on DBLP Social Network

Before presenting the random walk algorithms, we define the problems we are solving: given an author node $a \in A$, we compute a relevance score for each author $b \in A$. The result is a one-column vector containing all author scores with respect to a. We measure closeness of researchers so we can discover implicit communities in the DBLP data.

Recall that we extend the bipartite model into a directed bipartite graph $G' = (C', A, E')$, where A has m author nodes, C' is generated base on C and has $n * m$ nodes (we assume every node in C is split into m nodes). The basic intuition of our approach is to apply random walks on the adjacency matrix of graph G' starting from a given author node. To form the adjacency matrix, we first generate a matrix for directional edges from C' to A, which is $M_{(n*m) \times m}$, then form a matrix for edges from A to C', which is $N_{m \times (n*m)}$. In these two matrices, $M(\alpha, \beta)$ or $N(\alpha, \beta)$ indicates the weight of the directed edge from node α to node β in G' (0 means no such edge). A random walk starting from a node represented by row α in M (the same applies to N) takes the edge (α, β) based on the probability which is proportional to the edge weight over the sum of weight of all outgoing edges of α. Therefore, we normalize M and N such that every row sums up to 1. We can then construct the adjacency matrix J of G':

$$J_{(n*m+m) \times (m+n*m)} = \begin{pmatrix} 0 & (Norm(N))^T \\ (Norm(M))^T & 0 \end{pmatrix}$$

We then transform the given author node α into a one-column vector v_α consisting of $(n * m + m)$ elements. The value of the element corresponding to author α is set to 1. We now need to compute a steady-state vector u_α, which contains relevance scores of all nodes in the graph model. The scores for the author nodes are the last m elements of the vector. The result is achieved based on the following lemma and the RWR approach.

Lemma 1. *Let c be the probability of restarting random walk from node α. Then the steady-state vector u_α satisfies the following equation:*

$$u_\alpha = (1 - c)Ju_\alpha + cv_\alpha$$

See [22] for proof of the lemma.

Algorithm 1. The Random Walk with Restart Algorithm

Input: node $\alpha \in A$, bipartite graph model G, restarting probability c, converge threshold ϵ.
Output: relevance score vector A for author nodes.
1. Construct graph model G' for co-authorship based
   on G. Compute the adjacency matrix J of G'.
2. Initialize $v_\alpha = 0$.
   set value for α to 1: $v_\alpha(\alpha) = 1$.
3. While $(\Delta u_\alpha > \epsilon)$
   $$u_\alpha = (1 - c)\left(\frac{(Norm(N))^T u_{\alpha(n*m+1:n*m+m)}}{Norm(M)^T u_{\alpha(1:n*m)}}\right) + cv_\alpha$$
4. Set vector $A = u_{\alpha(n*m+1:n*m+m)}$
5. Return A.

Algorithm 1 applies the above lemma repeatedly until u_α converges. We set c to be 0.15 and ϵ to be 0.1, which gives the best convergence rate according to [23]. The bipartite structure of the graph model is used to save the computation of applying Lemma 1 in step 3. The last m elements of the result vector $u_{\alpha(n*m+1:n*m+m)}$ contains the relevance score for all author nodes in A.

We extend algorithm 1 for the tripartite graph model $G'' = (C, A, T, E'')$. Assume we have n conferences, m authors and l topics in G', we can represent all relations using three corresponding matrices: $U_{n\times m}$, $V_{m\times l}$ and $W_{n\times l}$. We normalize them such that every column sum up to 1: $Q(U) = col\_norm(U)$, $Q(U^T) = col\_norm(U^T)$. We then construct the adjacency matrices of G'' after normalization:

$$J_{CA} = \begin{pmatrix} 0 & Q(U) \\ Q(U^T) & 0 \end{pmatrix}$$

$$J_{CT} = \begin{pmatrix} 0 & Q(W) \\ Q(W^T) & 0 \end{pmatrix}$$

$$J_{AT} = \begin{pmatrix} 0 & Q(V) \\ Q(V^T) & 0 \end{pmatrix}$$

Similarly, given a node $\alpha \in C$, we want to compute a relevance score for all nodes that are in C, A, T. We apply the RWR approach following the visiting sequence until convergence, e.g., walk from author to conference, to topic, and back to author if we want to rank authors (see Algorithm 2). There are $m + n + l$ elements for all nodes in the graph model in the result relevance score vector. The value of the corresponding node, either starting author, topic or conference, is initialized to 1. After the random walk algorithm terminates, scores for conference, author and topic nodes are recorded from 1 to n, from $n + 1$ to $n + m$ and from $n + m + 1$ to $n + m + l$ in the vector, respectively.

Here we show a simple random walk on the conference-author network example we give in Table 1. The relational matrix M of the network is shown as follows.

$$M = \begin{pmatrix} & a\ b\ c\ d\ e \\ \hline VLDB & 5\ 1\ 0\ 0\ 0 \\ KDD & 3\ 0\ 7\ 0\ 7 \\ SIGMOD & 0\ 4\ 0\ 3\ 2 \end{pmatrix}$$

Then we build and normalize the adjacency matrix J of the graph shown in Figure 1.

$$J = \begin{pmatrix} & VLDB\ KDD\ SIGMOD\ a\ b\ c\ d\ e \\ \hline VLDB & 0 & 0 & 0 & 5\ 1\ 0\ 0\ 0 \\ KDD & 0 & 0 & 0 & 3\ 0\ 7\ 0\ 7 \\ SIGMOD & 0 & 0 & 0 & 0\ 4\ 0\ 3\ 2 \\ a & 5 & 3 & 0 & 0\ 0\ 0\ 0\ 0 \\ b & 1 & 0 & 4 & 0\ 0\ 0\ 0\ 0 \\ c & 0 & 7 & 0 & 0\ 0\ 0\ 0\ 0 \\ d & 0 & 0 & 3 & 0\ 0\ 0\ 0\ 0 \\ e & 0 & 7 & 2 & 0\ 0\ 0\ 0\ 0 \end{pmatrix}$$

$$J_{normalize} =$$

	VLDB	KDD	SIGMOD	a	b	c	d	e
VLDB	0	0	0	0.62	0.2	0	0	0
KDD	0	0	0	0.38	0	1.0	0	0.77
SIGMOD	0	0	0	0	0.8	0	1.0	0.22
a	0.84	0.18	0	0	0	0	0	0
b	0.16	0	0.44	0	0	0	0	0
c	0	0.41	0	0	0	0	0	0
d	0	0	0.33	0	0	0	0	0
e	0	0.41	0.22	0	0	0	0	0

A random walk on this graph moves from one node to one of its neighbours at random but the probability of picking a particular edge is proportional to the weight of the edge out of the sum of weights of all edges that connect to this node. For example, if we start from node SIGMOD, we build u as the start vector:

$$u = \{0, 0, 1, 0, 0, 0, 0, 0\}^T$$

After the first step of the first iteration, the random walk hits the author nodes with $b = 1 * 0.44, d = 1 * 0.33, e = 1 * 0.22$.

$$u = \{0, 0, 0, 0, 0.44, 0, 0.33, 0.22\}^T$$

In the next step of the first iteration, the chance that the random walk goes back to SIGMOD is $0.44 * 0.8 + 0.33 * 1 + 0.22 * 0.22 = 0.73$. The other 0.27 goes to the other two conference nodes.

$$u = \{0.09, 0.18, 0.73, 0, 0, 0, 0, 0\}^T$$

The vector will converge after a few iterations and gives a stable score to every node, which is the probability of a random walk may hit this node. However, the fact that these scores are always the same no matter where the walk begins makes the approach incapable for ranking for different given starting points. This problem can be solved by

---

Algorithm 2. Random Walk Algorithm for Tripartite Model

Input: node α, tripartite graph model G'', restarting probability c, converge threshold ϵ.
Output: relevance score vector c, a and t for author, conference and topic nodes.
1. Compute the adjacency matrices J_{CA}, J_{CT} and J_{AT} of G''.
2. Initialize $v_\alpha = 0$, set element for α to 1: $v_\alpha(\alpha) = 1$.
3. While ($\Delta u_\alpha > \epsilon$)
   $u_{\alpha(n+1:n+m)} = (Q(U^T) * u_{\alpha(1:n)})$
   $u_{\alpha(n+m+1:n+m+l)} = (Q(V^T) * u_{\alpha(n+1:n+m)})$
   $u_{\alpha(1:n)} = (Q(W) * u_{\alpha(n+m+1:n+m+l)})$
   $u_\alpha = (1 - c)u_\alpha + cv_\alpha$
4. Set vector $c = u_{\alpha(1:n)}$, $a = u_{\alpha(n+1:n+m)}$,
   $t = u_{\alpha(n+m+1:n+m+l)}$.
6. Return c, a, t.

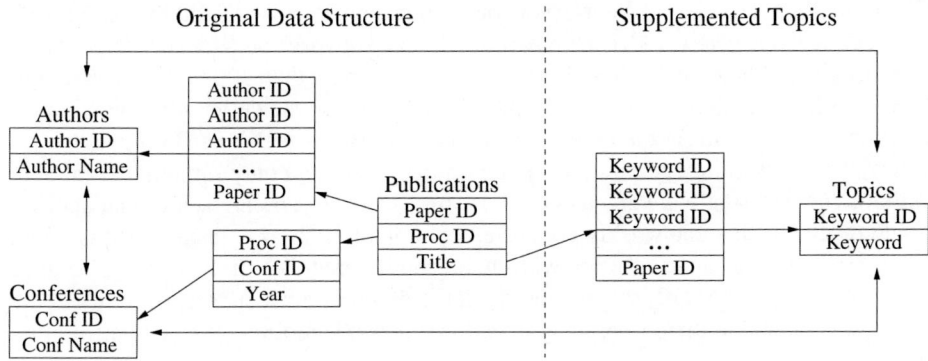

Fig. 3. Our Data Structure extracted from DBLP and Citeseer

random walk with restart: in each random walk iteration, the walker goes back to the start node with a restart probability. Therefore, nodes that are closer to the starting node now have a higher chance to be visited and obtain larger ranking score.

4 Exploring DBLP Communities

In the academic world, since a researcher could usually belong to multiple related communities, e.g., Database and AI, it is unnecessary and improper to classify this researcher into any specific arbitrary communities. Therefore, in our experiment, we focus on investigating the closeness of researchers, i.e., we are interested at *how* and *why* two people are in the same community, instead of *which* community they are in.

4.1 DBLP Database

We downloaded the publication data for conferences from the DBLP website[9] in July 2007. Any publication after that date is not included in our experimental data. Moreover, we kept only conference proceedings and removed all journals and other publications. These were minimal compared to the conference publications. The data structure is shown in Figure 3. We extracted topics based on keyword frequency from paper titles in DBLP data and abstracts from Citeseer[10], which provides abstracts of about 10% of all papers. First we manually selected a list of stopwords to remove frequently used but non-topic-related words, e.g., "Towards", "Understanding", "Approach", etc. Then we counted frequency of every co-located pairs of stemmed words and selected the top 1000 most frequent bi-grams as topics. Additionally, we manually added several tri-grams, e.g. World Wide Web, Support Vector Machine, etc., since we observe both bilateral bi-grams to be frequent (e.g. World Wide and Wide Web). We chose to use bi-grams because they can distinguish most of the research topics, e.g, Relational

---

[9] http://www.informatik.uni-trier.de/~ley/db/
[10] http://citeseer.ist.psu.edu/

Database, Web Service and Neural Network, while single keywords fail to separate different topics, e.g. "Network" can be part of "Social Network" or "Network Security".

Since the publication database is huge (it contains more than 300,000 authors, about 3,000 conferences and the selected 1,000 N-gram topics), the entire adjacency matrix becomes too big to make the random walk efficient. However, we can compute the result by first performing graph partitioning on the model and only running the random walk on the part where the given author is. This approach can only achieve an approximate result, since some weakly connected communities are separated, but it is much faster since we end-up computing with much smaller matrices. In this paper, we used the METIS algorithm [10] to partition the large graph into ten subgraphs of about the same size. Note that the proposed approach is independent of the selected partitioning method.

4.2 The DBconnect System

After the author-conference-topic data extraction from the DBLP database, we generate lists of people with high relevance scores with respect to different given researchers. Our ongoing project *DBconnect*, which is a navigational system to investigate the community connections and relationships, is built to explore the result lists of our random walk approach on the academic social network. An online demo for DBConnect is available at[11]. Figure 4 shows a screenshot of the author interface of our *DBconnect* system. There are eight lists displayed for a given author in the current version. Clicking on any of the hyper-linked names will generate a page with respect to that selected entity. We explain details of each of the lists below.

- *Academic Information*
  Academic statistics for the given author are shown in this list, which contain three components: conference contribution, earliest publication year and average publication per year are extracted from DBLP; the H-index [27] is calculated based on information retrieved from Google Scholar[12]; approximate citation numbers are retrieved from Citeseer[13]. The query terms for Google Scholar and Citeseer are automatically generated based on the author names. Users can submit an alternative query which gives a more accurate result from the search engines. We also provide a visualization of the H-index. One can click the "See graph" link beside the H-index numbers. Figure 5 shows an example of H-index visualization.
- *Related Conferences*
  This list is generated by the random walk, which starts from the given author, on an author-conference-topic model and is ordered by their relevance score, in descending order. These are not necessarily the conferences where the given researcher published but the conferences related to the topics and authors that are also related to the reference researcher. Clicking on the conference name leads to a new page with topics and authors related to the chosen conference.

---

[11] http://kingman.cs.ualberta.ca/research/demos/content/dbconnect/
[12] http://scholar.google.com/
[13] http://citeseer.ist.psu.edu/

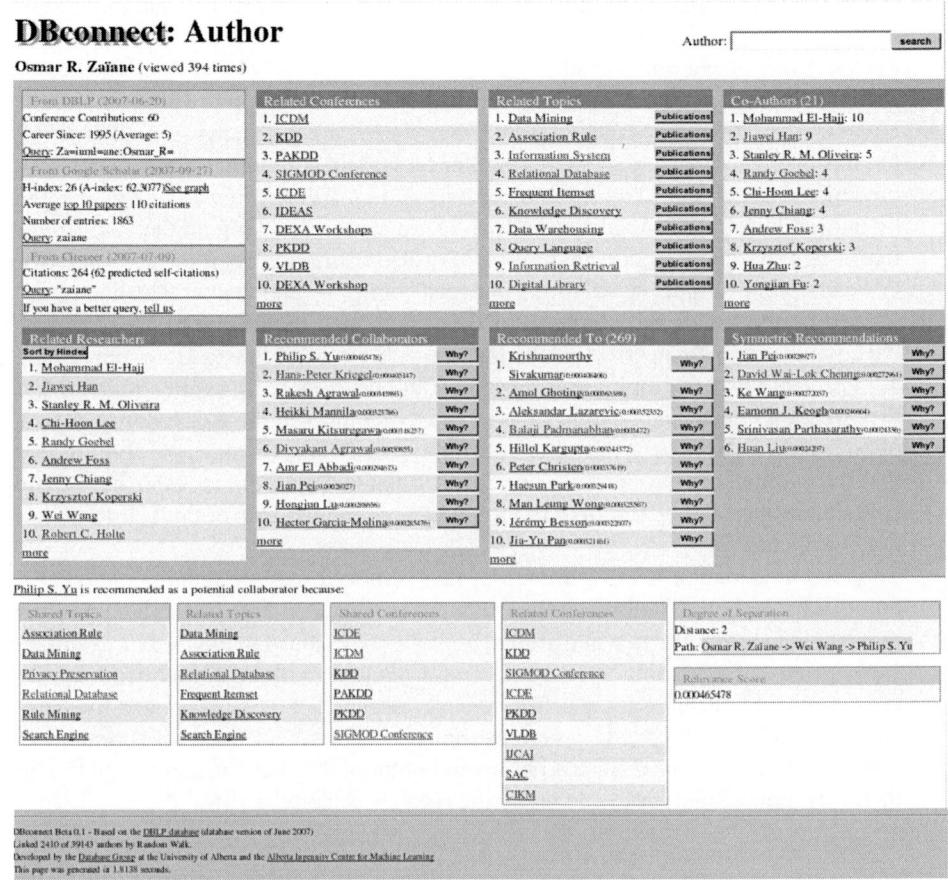

Fig. 4. DBconnect Interface Screenshot for an author

– *Related Topics*
  This list is ordered by the relevance scores from a random walk on the tripartite model. Clicking on the button "Publications" after each topic provides the papers that the given author has published on that topic, i.e. the papers of the given author that contains the N-gram keywords in their titles or abstracts. Similarly, these are not necessarily the topics that the given author has worked on, but the topics most related to his topics, attended conferences and colleagues.

– *Co-authors*
  The co-author list reports the number of publications that different researchers co-authored with the given person.

– *Related Researchers*
  This list is based on the bipartite graph model with only conference and author entities, i.e. we apply our extended bipartite graph model to emphasize the co-authorship. The result implies that the given author is related to the same conferences and via the same co-authors as these listed researchers. In most cases,

most related researchers to the given author are co-authors and co-authors of co-authors.

– *Recommended Collaborators*

This list is based on the tripartite graph author-conference-topic. Since co-authors are treated as "observed collaborators", their names are not shown here. The result implies that the given author shares similar topics and conference experiences with these listed researchers, hence the recommendation. The relevance score calculated by our random walk is displayed following the names. Clicking on the "why" button brings the detailed information of the relationship between the two authors. For example, in Figure 4, relations between Philip Yu and Osmar Zaïane are described by the topics and conferences they share, and the degree of separation in the co-authorship chain ($A \rightarrow B$ means A and B are co-authors). Here, the "Share Topics" table lists the topics that these two authors both have publications on and the "Related Topics" table shows the topics that appear in the Related Topics lists of both authors. Similarly, the "Shared Conferences" table displays the conferences that the two authors have attended and the "Related Conferences" table shows the conferences that can be found in the Related Conferences lists of both authors.

– *Recommended To*

The recommendation is not symmetric, i.e., author A may be recommended as a possible future collaborator to author B but not vice versa. This phenomenon is due to the unbalanced influence of different authors in the social network. For example, Jiawei Han has a significant influence with his 196 conference publications, 84 co-authors and H-index 63. He has been recommended as collaborator for 6201 authors, but apparently only a few of them is recommended as collaborators to him. The Recommended To list shows the authors that have the given author in their recommendation list, ordered by the relevance score.

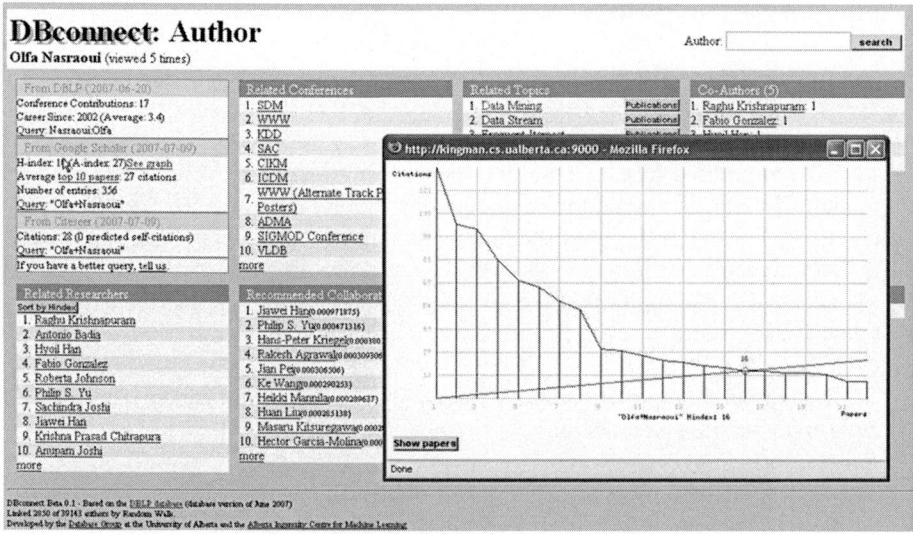

Fig. 5. DBconnect Interface Screenshot for H-Index Visualization

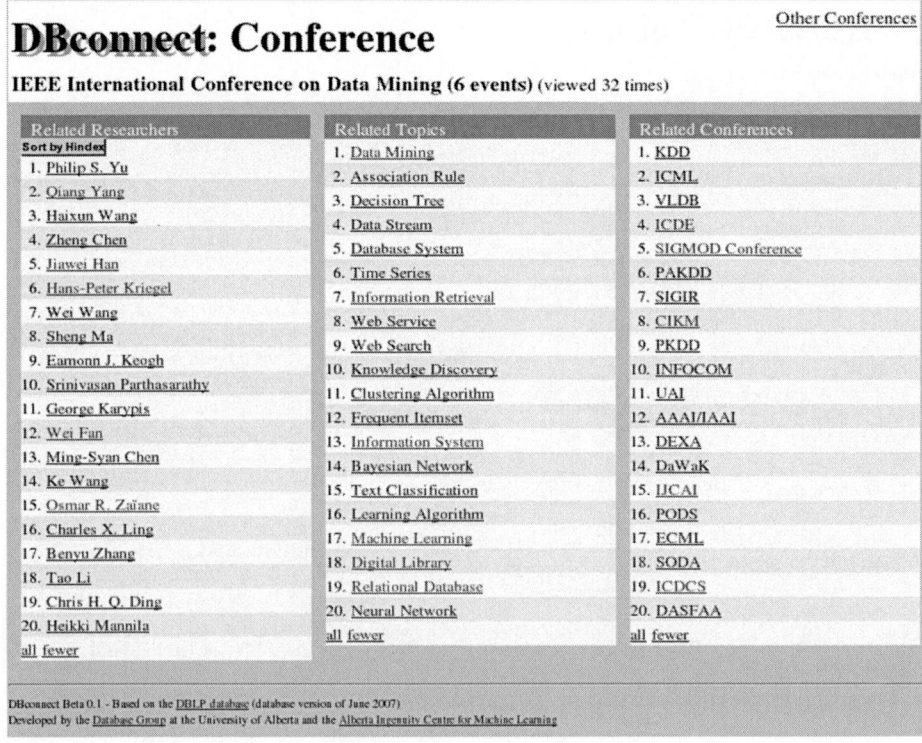

Fig. 6. DBconnect Interface Screenshot for conference ICDM

- *Symmetric Recommendations*
  This list shows the authors that have been recommended to the given author and have the given author on their recommendation list.

Note that while there is some overlap between the list of related researchers and the list of recommended collaborators, there is a fundamental difference and the difference by no means implies that collaboration with the missing related researchers is discouraged. They are simply two different communities in the network even though they overlap. The list of related researchers is obtained from relationships derived from co-authorships and conferences by a RWR on an extended bipartite graph with co-authorship relations. The result is a quasi-obvious list due to the closeness from co-authors. This list could create a sort of trust in the system given the clear closeness of this community. The list of recommended collaborators could be perceived as a more distant community and thus as an interesting discovery. It is obtained without co-authorship but with relations from topics. We use a RWR on a tripartite graph authors/conferences/topics. The explanation on the why collaborators are recommended (i.e. common conferences and topics, and degree of separation) establishes more trust in the recommendation. A systematic validation of these lists is difficult but the cases we manually substantiated were satisfactory and convincing.

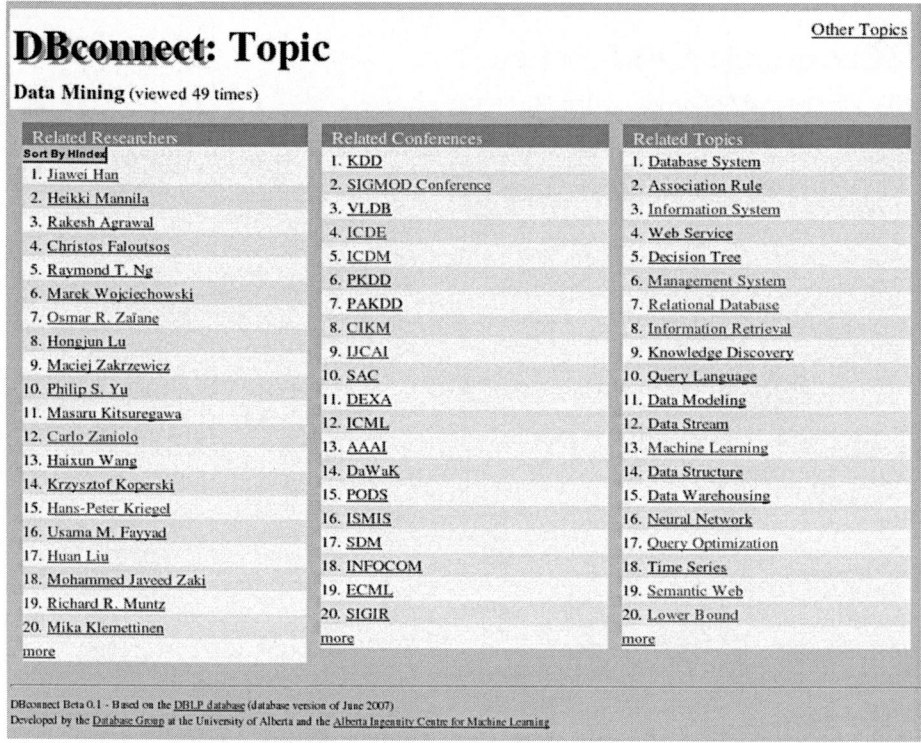

Fig. 7. DBconnect Interface Screenshot for topic Data Mining

Clicking on any conference name shows a conference page. Figure 6 illustrates an example when the entity "ICDM" is selected. Conferences have their own related conferences, authors and topics. Note that the topics here mean the most frequent topics used within titles and abstracts of papers published in the given conference.

Clicking on the topics leads to a new page with conferences, authors and topics related to the chosen topic. Note again that this relationship to topics comes from paper titles and abstracts. Figure 7 shows an example when the topic "Data Mining" is selected.

5   Conclusions and Future Work

In this paper, we extend a bipartite graph model to incorporate co-authorship, and propose a random walk approach to find related conferences, authors, and topics for a given entity. The main idea is to use a random walk with restarts on the bipartite or tripartite model of DBLP data to measure the closeness between any two entities. The result, the relevance score, can be used to understand the relationship between entities and discover the community structure of the corresponding data. We basically use the relevance score to rank entities based on importance given a relationship.

We also present our ongoing work DBconnect, which can help explore the relational structure and discover implicit knowledge within the DBLP data collection. Not all of the more than 360,000 authors are indexed in DBconnect at the time of printing as the random walks are time consuming. A queue of authors is continuously processed in parallel and authors can be prioritized in the queue by request.

The work we presented in this paper is still preliminary. We have implemented a prototype[14]. However, more work is needed to verify the value of the approach. The lists of related conferences, topics and researchers to a given author are interesting and can be used to help understand the entity closeness and research communities. While the output of DBconnect is satisfactory and the manual substantiation confirms acceptable and suitable lists (as opposed to lists provided by DBLife), some systematic evaluation is still desired. However, validation of the random walk is difficult and we are considering devising methods to confirm the accuracy of the relevance score and the generated lists. Moreover, it is hard to extract correct topics for researchers since the only available information is the title of the paper, which usually does not suffice to describe the content. Some titles are even unconventionally unrelated to the content of the paper only to attract attention or are metaphoric. We are considering implementing a hierarchy of topics to group similar topics and ease the browsing of the long list of related topics in computer science. We also plan to address the issue of acronyms in titles that are currently discarded. For example HMM for Hidden Markov Model is currently eliminated due to infrequency while relevant as a topic. In addition, the matrix multiplications in the random walk process make it expensive to compute. Improving the efficiency of the random walk without jeopardizing its effectiveness is necessary since the computations for relevance score estimation need to be redone continuously as the the DBLP database never ceases to grow.

Acknowledgments

Our work is supported by the Canadian Natural Sciences and Engineering Research Council (NSERC), the Alberta Ingenuity Centre for Machine Learning (AICML), and the Alberta Informatics Circle of Research Excellence (iCORE).

References

1. Brin, S., Page, L.: The anatomy of a large-scale hypertextual web search engine. In: Seventh International World Wide Web Conference, Brisbane, Australia, pp. 107–117 (1998)
2. Buchanan, M.: Nexus: Small worlds and the groundbreaking theory of networks. W. W. Company, Inc., Norton (2003)
3. DBLP (Digital Bibliography & Library Project) Bibliography database, http://www.informatik.uni-trier.de/~ley/db/
4. Doan, A., Ramakrishnan, R., Chen, F., DeRose, P., Lee, Y., McCann, R., Sayyadian, M., Shen, W.: Community information management. IEEE Data Engineering Bulletin, Special Issue on Probabilistic Databases 29(1) (2006)
5. Girvan, M., Newman, M.E.J.: Community structure in social and biological networks. Proceedings of the National Academy of Science USA 99, 8271–8276 (2002)

---

[14] http://kingman.cs.ualberta.ca/research/demos/content/dbconnect/

6. Haveliwala, T.H.: Topic-sensitive pagerank. In: WWW: Proceedings of the 11th international conference on World Wide Web, pp. 517–526 (2002)
7. He, J., Li, M., Zhang, H.-J., Tong, H., Zhang, C.: Manifold-ranking based image retrieval. In: MULTIMEDIA: Proceedings of the 12th annual ACM international conference on Multimedia, pp. 9–16 (2004)
8. Holme, P., Huss, M., Jeong, H.: Subnetwork hierarchies of biochemical pathways. Bioinformatics 19, 532–538 (2003)
9. Jeh, G., Widom, J.: Simrank: a measure of structural-context similarity. In: KDD (2002)
10. Karypis, G., Kumar, V.: Multilevel k-way partitioning scheme for irregular graphs. Journal of Parallel and Distriuted Computing 48(1), 96–129 (1998)
11. Kernighan, B.W., Lin, S.: An efficient heuristic procedure for partitioning graphs. Bell System Technical Journal 49, 291–307 (1970)
12. Klink, S., Reuther, P., Weber, A., Walter, B., Ley, M.: Analysing social networks within bibliographical data. In: Bressan, S., Küng, J., Wagner, R. (eds.) DEXA 2006. LNCS, vol. 4080, pp. 234–243. Springer, Heidelberg (2006)
13. Ley, M.: The DBLP computer science bibliography: Evolution, research issues, perspectives. In: Laender, A.H.F., Oliveira, A.L. (eds.) SPIRE 2002. LNCS, vol. 2476, pp. 1–10. Springer, Heidelberg (2002)
14. César Cazella, S., Campos Alvares, L.O.: An architecture based on multi-agent system and data mining for recommending research papers and researchers. In: Proc. of the 18th International Conference on Software Engineering and Knowledge Engineering (SEKE), pp. 67–72 (2006)
15. Nascimento, M.A., Sander, J., Pound, J.: Analysis of sigmod's co-authorship graph. SIGMOD Record 32(2), 57–58 (2003)
16. Newman, M.E.J.: The structure and function of complex networks. SIAM Review 45(2), 167–256 (2003)
17. Page, L., Brin, S., Motwani, R., Winograd, T.: The pagerank citation ranking: Bringing order to the web. Technical report, Stanford University Database Group (1998)
18. Pan, J.-Y., Yang, H.-J., Faloutsos, C., Duygulu, P.: Automatic multimedia cross-modal correlation discovery. In: KDD, pp. 653–658 (2004)
19. Pothen, A., Simon, H., Liou, K.P.: Partitioning sparse matrices with eigenvectorsof graphs. SIAM J. Matrix Anal. Appl. 11, 430–452 (1990)
20. Radicchi, F., Castellano, C., Cecconi, F., Loreto, V., Parisi, D.: Defining and identifying communities in networks. Proc. Natl. Acad. Sci. USA 101, 2658 (2004)
21. Smeaton, A.F., Keogh, G., Gurrin, C., McDonald, K., Sodring, T.: Analysis of papers from twenty-five years of sigir conferences: What have we been doing for the last quarter of a century. SIGIR Forum 36(2), 39–43 (2002)
22. Strang, G.: Introduction to linear algebra, 3rd edn. Wellesley-Cambridge Press (1998)
23. Sun, J., Qu, H., Chakrabarti, D., Faloutsos, C.: Neighborhood formation and anomaly detection in bipartite graphs. In: ICDM, pp. 418–425 (2005)
24. Tong, H., Faloutsos, C., Pan, J.-Y.: Fast random walk with restart and its applications. In: ICDM, pp. 613–622 (2006)
25. Tyler, J.R., Wilkinson, D.M., Huberman, B.A.: Email as spectroscopy: automated discovery of community structure within organizations. Communities and technologies, pp. 81–96 (2003)
26. Wasserman, S., Faust, K.: Social network analysis: Methods and applications. Cambridge University Press, Cambridge (1994)
27. Wendl, M.C.: H-index: however ranked, citations need context. Nature 449(403) (2007)
28. Yin, X., Han, J., Yu, P.S.: Linkclus: efficient clustering via heterogeneous semantic links. In: VLDB, pp. 427–438 (2006)

Dynamics of a Collaborative Rating System

Kristina Lerman

USC Information Sciences Institute, Marina del Rey, California 90292
lerman@isi.edu

Abstract. The rise of social media sites, such as blogs, wikis, Digg and
Flickr among others, underscores a transformation of the Web to a participatory medium in which users are actively creating, evaluating and
distributing information. The social news aggregator Digg allows users
to submit links to and vote on news stories. Like other social media sites,
Digg also allows users to designate others as "friends" and easily track
friends' activities: what new stories they submitted, commented on or
liked. Each day Digg selects a handful of stories to feature on its front
page. Rather than rely on the opinion of a few editors, Digg aggregates
opinions of thousands of its users to decide which stories to promote to
the front page. We construct two mathematical models of collaborative
decision-making on Digg. First, we study how collective rating of news
stories emerges from the decisions made by many users. The model takes
into account the effect that decisions made by a user's friends have on
the user. We also study how user's influence, as measured by her rank,
changes in time as she submits new stories and is befriended by other
Digg users. Solutions of both models reproduce the observed dynamics
of voting and user rank on Digg.

The Digg model that enables users to collectively rate news stories can be
generalized to the collaborative evaluation of document (or information)
quality. Mathematical analysis can be used as a tool to explore different
collaborative decision-making algorithms to select the most effective one
before the algorithm is ever implemented in a real system.

Keywords: News aggregation, social networks, dynamics, collaborative
rating, mathematical analysis.

1 Introduction

The new social media sites — blogs, wikis, MySpace, Flickr, del.icio.us, and their
ilk — have enjoyed phenomenal success in recent years. The extraordinary rise in
their popularity is underscoring a transformation of the Web to a participatory
medium where the users are actively creating, evaluating and distributing information. These sites share four elements: (1) users create or contribute content,
(2) users annotate content with tags, (3) users evaluate content and (4) users create social networks by designating other users with similar interests as friends or
contacts. These innovations help users solve hard information processing problems collaboratively, *e.g.*, detect public opinion trends in the blogosphere [1],

H. Zhang et al. (Eds.): WebKDD/SNA-KDD 2007, LNCS 5439, pp. 77–96, 2009.

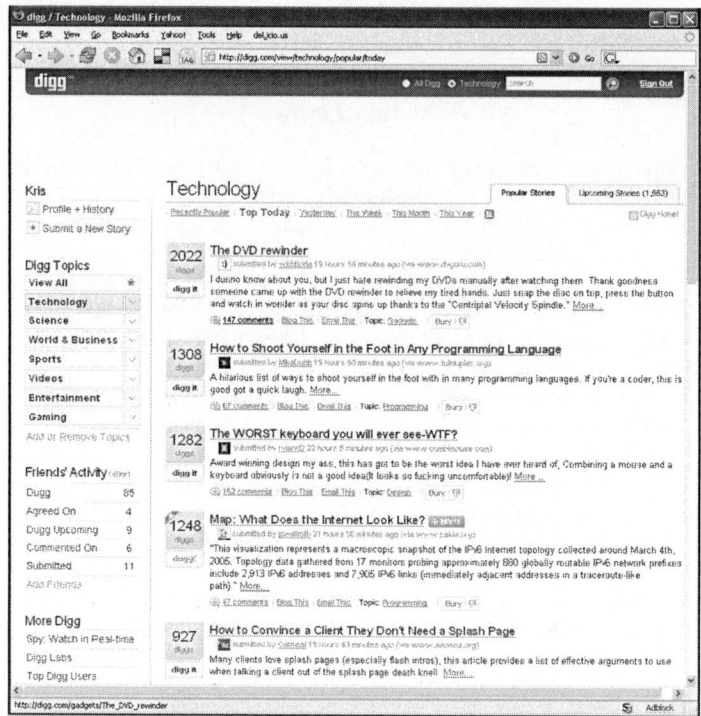

Fig. 1. Digg front page showing the technology section

construct taxonomies from the distributed tagging activities of many individuals [11], and use social networks as recommendation [5] and browsing aides [7].

One of the outstanding problems in information processing is how to evaluate the quality of documents or information in general. This problem crops up daily in information retrieval and Web search, where the goal is to find the information among the terabytes of data accessible online that is most relevant to a user's query. The standard practice of search engines is to identify all documents that contain user's search terms and rank results. Google revolutionized Web search by exploiting the link structure of the Web — created through independent activities of many Web page authors — to evaluate the contents of information on Web pages. Google, therefore, relies on an implicit rating scheme, where a link to a document is interpreted as a vote for it. Best seller lists are another example of an implicit rating system. An alternative to this is explicit rating, where a user assigns a rating, or a positive (or negative) vote to some document. Explicit ratings are used in many commercial collaborative filtering applications, on Amazon and Netflix for example, to recommend new products to users.

Social news aggregators like Digg[1] became popular in part because they rely on the distributed opinions of many independent voters to help users find the

---
[1] http://digg.com

most interesting news stories. The functionality of Digg is very simple: users submit stories they find online, and other users rate them by voting on them. Each day Digg selects a handful of stories to feature on its front pages. Although the exact formula for how a story is selected for the front page is kept secret, so as to prevent users from "gaming the system" to promote fake stories or spam, it appears to take into account the number of votes a story receives. The promotion mechanism, therefore, does not depend on the decisions of a few editors, but emerges from the activities of many users. This type of collective decision making can be extremely efficient, outperforming special-purpose algorithms. For example, the news of Rumsfeld's resignation in the wake of the 2006 U.S. Congressional elections broke Digg's front page within 3 minutes of submission, 20 minutes before Google News showed it [13].

Designing a complex system like Digg, that exploits the emergent behavior of many independent evaluators, is exceedingly difficult. The choices made in the user interface, e.g., whether to allow users to see stories their friends voted on or the most popular stories within the past week or month, can have a dramatic impact on the behavior of the system and on user experience. Outside of running the system or perhaps simulating it, designers have little choice in how they evaluate the performance of different designs. Mathematical analysis can be used as a tool to explore the design space of collaborative rating algorithms to find parameters that optimize a given set of metrics (story timeliness vs interest, etc.), or eliminate unintended artifacts, before the algorithms are ever implemented in a real system.

This paper studies collaborative rating of news stories on Digg. Although Digg focuses on news stories and blogs, its collaborative rating approach can be extended to evaluating other kinds of information. We present a mathematical model of the dynamics of collective voting on Digg and show that solutions of the model strongly resemble votes received by actual news stories. By submitting and voting on stories, Digg users are also ranked by Digg. We present a second model that describes how a user's rank changes in time. We show that this model appears to explain the observed behavior of user rank.

The paper is organized as follows: In Section 2 we describe Digg's functionality and features in detail. In Section 3 we develop a model of collective voting. We compare solutions of the model to the behavior of actual stories. In Section 4 we develop a model of the dynamics of user rank and compare its solutions to the observed changes in user rank. In Section 5 we discuss limitations of mathematical modeling and identify new directions.

2   Anatomy of Digg

Digg is a social news aggregator that relies on users to submit and moderate stories. A typical Digg page is shown in Figure 1. When a story is submitted, it goes to the upcoming stories queue. There are 1-2 new submissions every minute. They are displayed in reverse chronological order of being submitted, 15 stories to the page, with the most recent story at the top. The story's title is a link to the source, while clicking on the number of diggs takes one to the page describing the story's activity on Digg: the discussion around it, the list of people who voted on it, etc.

When a story gets enough votes, it is promoted to the front page. The vast majority of people who visit Digg daily, or subscribe to its RSS feeds, read only the front page stories; hence, getting to the front page greatly increases a story's visibility. Although the exact promotion mechanism is kept secret and changes periodically, it appears to take into account the number of votes the story receives. Digg's front page, therefore, emerges by consensus between many independent users.

Digg allows users to designate others as friends and makes it easy to track friends' activities.[2] The left column of the front page in Figure 1 summarizes the number of stories friends have submitted, commented on or liked (dugg) recently. Tracking activities of friends is a common feature of many social media sites and is one of their major draws. It offers a new paradigm for interacting with information — *social navigation and filtering*. Rather than actively searching for new interesting content, or subscribing to a set of predefined topics, users can now put others to task of finding and filtering information for them.

Top users list. Until February 2007 Digg ranked users according to how many of the stories the user submitted were promoted to the front page. Clicking on the Top Users link allowed one to browse through the ranked list of users. There is speculation that ranking increased competition, leading some users to be more active in order to improve their position on the Top users list. Digg discontinued making the list publicly available, citing concerns that marketers were paying top users to promote their products and services [15], although it is now available through a third party.[3]

Social recommendation. The Friends interface allows Digg users to see the stories their friends submitted or liked recently; therefore, it acts as a social recommendation system. By comparing users who voted on the story with the social network of the submitter, we showed that users tend to like (and vote on) the stories their friends submit and to a lesser extent, they tend to like the stories their friends like [5]. Thus, social networks on Digg contribute to how successful a user is at getting her stories promoted to the front page. A user's success rate is defined as the fraction of the stories the user has submitted that have been promoted to the front page. We used statistics about the activities of the top 1020 users to show that users with bigger social networks are more successful at getting their stories promoted [5].

---

[2] Note that the friend relationship is asymmetric. When user A lists user B as a *friend*, user A is able to watch the activity of B but not vice versa. We call A the *reverse friend* of B.

[3] http://www.efinke.com/digg/topusers.html

3   Dynamics of Ratings

In order to study how the front page emerges from independent opinions of many users, we tracked both the upcoming and front page stories in Digg's technology section. We collected data by scraping Digg site with the help of Web wrappers, created using tools provided by Fetch Technologies[4]:

digg-frontpage wrapper extracts a list of stories from the first 14 front pages. For each story, it extracts submitter's name, story title, time submitted, number of votes and comments the story received.

digg-all wrapper extracts a list of stories from the first 20 pages in the upcoming stories queue. For each story, it extracts the submitter's name, story title, time submitted, number of votes and comments the story received.

top-users wrapper extracts information about the top 1020 of the recently active users. For each user, it extracts the number of stories the user has submitted, commented and voted on; number of stories promoted to the front page; users's rank; the list of friends, as well as reverse friends or "people who have befriended this user."

Digg-frontpage and *digg-all* wrappers were executed hourly over a period of a week in May and in July 2006.

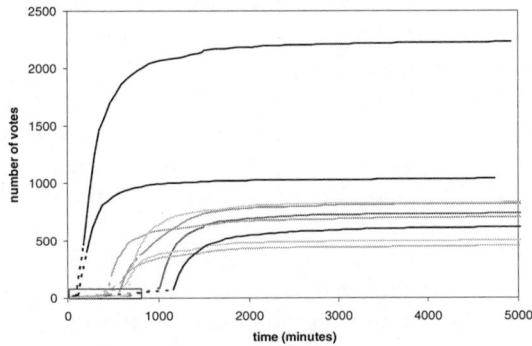

Fig. 2. Dynamics of votes of select stories over a period of four days. The small rectangle in the lower corner highlights votes received by stories while in the upcoming stories queue. Dashes indicate story's transition to the front page.

We identified stories that were submitted to Digg over the course of approximately one day and followed them over a period of several days. Of the 2858 stories submitted by 1570 distinct users, only 98 stories by 60 users made it to the front page. Figure 2 shows evolution of the ratings (number of votes) of select stories. The basic dynamics of all stories appears the same: while in the upcoming queue, a story accrues votes at some slow rate, and once promoted

---

[4] http://fetch.com/

to the front page, it accumulates votes at a much faster rate. As the story ages, accumulation of new votes slows down, and the story's rating saturates at some value. This value depends on how *interesting* the story is to the Digg community.

It is worth noting that the top-ranked users are not submitting the most interesting stories (that get the most votes). Slightly more than half of the stories our data set came from 14 top-ranked users (rank< 25) and 48 stories came from 45 low-ranked users. The average "interestingness" of the stories submitted by the top-ranked users is 600, almost half the average "interestingness" of the stories submitted by low-ranked users. A second observation is that top-ranked users are responsible for multiple front page stories. A look at the statistics about top users provided by Digg shows that this is generally the case: of the more than 15,000 front page stories submitted by the top 1020 users, the top 3% of the users are responsible for 35% of the stories. This can be explained by social recommendation and the observation that top users have bigger, more active social networks.

3.1 Mathematical Model

Our goal is not only to produce a mathematical model that can explain — and predict — the dynamics of collective voting on Digg, but one that can also be used as a tool to study the emergent behavior of collaborative rating systems. Our modeling approach is motivated by the stochastic processes-based framework we developed to study collective behavior of multi-agent systems [9,8,6]. We view an agent, be it a software system, a robot or a Digg user, as an automaton that takes certain actions based on external conditions, and if necessary, its internal state. Although human behavior is certainly very complex, in the prescribed environment of Digg, users tend to follow a simple set of rules: (*1*) when a user sees a story, she will vote for it with some probability; (*2*) a user will befriend another user with some probability when she sees that user's name on the front page or the Top Users list; (*3*) a user will submit new stories to Digg at some rate.

Below we present a mathematical model that describes how the number of votes, m, a story receives changes over time t. This model is based on rule (*1*) above. The model of rank dynamics presented in the next section is based on the last two rules.

We parameterize a story by its *interestingness* coefficient r, which simply gives the probability that a story will receive a (positive) vote once seen. This is an *ad hoc* parameter to characterize how relevant or interesting a story is to Digg audience. The number of votes a story receives depends on its *visibility*, which simply means how many people can see a story and follow the link to it. The factors that contribute to the story's visibility include:

- visibility on the front page
- visibility in the upcoming stories queue
- visibility through the Friends interface

Digg offers additional ways to see popular stories: e.g., most popular stories submitted over the preceding week or month. We assume that these browsing modalities do not generate significant views, and focus on the simpler model that takes into account solely the factors enumerated above.

We parameterize a Digg user by her social network or the number of reverse friends, S, she has. These are the users who are watching her activity.

Model Parameters. We use a simple threshold to model how a story is promoted to the front page. When the number of votes a story receives is fewer than h, the story is visible in the upcoming queue; when $m \geq h$, the story is visible on the front page. This seems to approximate Digg's promotion algorithm as of May 2006: in our data set we did not see any front page stories with fewer than 44 votes, nor upcoming stories with more than 42 votes.

Visibility on front page. A story's visibility on the front page decreases as newly promoted stories push it farther down the list. While we do not have data about Digg visitors' behavior, specifically, how many visit Digg and proceed to page 2, 3 and so on, we propose to describe it by a simple model that holds that some fraction c_f of the visitors to the current front page proceed to the next page. Thus, if N users visit Digg's front page within some time interval, $c_f N$ users see the second page stories, and $c_f^{p-1} N$ users see page p stories.

Visibility on the upcoming stories queue. A similar model describes how a story's visibility in the upcoming stories queue decreases as it is pushed farther down the list by the newer submissions. If a fraction c of Digg visitors proceed to the upcoming stories section, and of these, a fraction c_u proceed to the next upcoming page, then $cc_u N$ of Digg visitors see second page stories, and $cc_u^{q-1} N$ users see page q stories.

Figure 3(a) shows how the current page number of a story on the front page, p, and the upcoming queue, q, changes in time for three randomly chosen stories from the May data set. The data is fit well by lines $\{p, q\} = k_{\{u,f\}}t$ with slopes $k_u = 0.060$ pages/m (3.60 pages/hr) for the upcoming stories and $k_f = 0.003$ pages/m (0.18 pages/hr) for the front page stories.

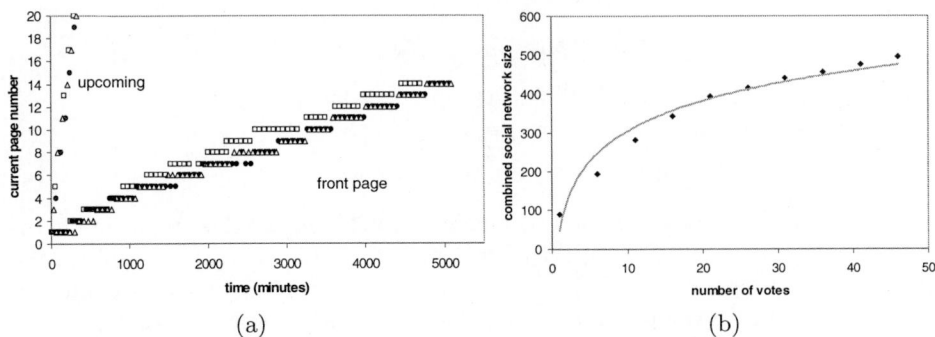

(a)   (b)

Fig. 3. Parameter estimation from data. (a) Current page number of a story on the upcoming stories queue and the front page vs time for three different stories. (b) Growth of the combined social network of the first 46 users to vote on a story.

Visibility through the Friends interface. The Friends interface offers the user ability to see the stories his friends have (i) submitted, (ii) liked (voted on), (iii) commented on during the preceding 48 hours or (iv) friends' stories that are still in the upcoming stories queue. Although it is likely that users are taking advantage of all four features, we will consider only the first two in the analysis. These features closely approximate the functionality offered by other social media sites: for example, Flickr allows users to see the latest images his friends uploaded, as well as the images a friend liked (marked as favorite). We believe that these features are more familiar to the user and used more frequently than the other features.

Friends of the submitter: S is the number of reverse friends the story's submitter has. These are the users who are watching the submitter's activities. We assume that these users visit Digg daily, and since they are likely to be geographically distributed across many time zones, they see the new story at an hourly rate of $a = S/24$. The story's visibility through the submitter's social network is therefore $v_s = a\Theta(S - at)\Theta(48 - t)$. $\Theta(x)$ is a step function whose value is 1 when $x \geq 0$ and 0 when $x < 0$. The first step function accounts for the fact that the pool of reverse friends is finite. As users from this pool read the story, the number of potential readers gets smaller. The second function accounts for the fact that the story will be visible through the Friends interface for 48 hours after submission only.

Friends of the voters: As the story is voted on, it becomes visible to more users through the see the "stories my friends dugg" part of the Friends interface. Figure 3(b) shows how S_m, the combined social network of the first m users to digg the story, changes as the story gets more votes. Although S_m is highly variable from story to story, it's average value (over 195 stories) has consistent growth: $S_m = 112.0 * log(m) + 47.0$. The story's visibility through the friends of voters is $v_m = bS_m\Theta(h - m)\Theta(48hrs - t)$, where b is a scaling factor that depends on the length of the time interval: for hourly counts, it is $b = 1/24$.

Dynamical Model. In summary, the four factors that contribute to a story's visibility are:

$$v_f = c_f^{p(t)-1} N\Theta(m(t) - h) \tag{1}$$

$$v_u = cc_u^{q(t)-1} N\Theta(h - m(t))\Theta(24hrs - t) \tag{2}$$

$$v_s = a\Theta(S - at)\Theta(48hrs - t) \tag{3}$$

$$v_m = bS_m\Theta(h - m(t))\Theta(48hrs - t) \tag{4}$$

t is time since the story's submission. The first step function in v_f and v_u indicates that when a story has fewer votes than required for promotion, it is visible in the upcoming stories pages; otherwise, it is visible on the front page. The second step function in the v_u term accounts for the fact that a story stays in the upcoming queue for 24 hours only, while step functions in v_s and v_m model the fact that it is visible in the Friends interface for 48 hours. The story's current page number on the upcoming stories queue q and the front page p change in time according to:

$$p(t) = (k_f(t - T_h) + 1)\Theta(T_h - t) \tag{5}$$

$$q(t) = k_u t + 1 \tag{6}$$

with $k_u = 0.060$ pages/min and $k_f = 0.003$ pages/min. T_h is the time the story is promoted to the front page.

The change in the number of votes m a story receives during a time interval Δt is

$$\Delta m(t) = r(v_f + v_u + v_s + v_m)\Delta t . \tag{7}$$

3.2 Solutions

We solve Equation 7 subject to the initial conditions $m(t = 0) = 1$, $q(t = 0) = 1$, as it starts with a single vote coming from the submitter himself. The initial condition for the front page is $p(t < T_h) = 0$, where T_h is the time the story was promoted to the front page. We take Δt to be one minute. The solutions of Equation 7 show how the number of votes received by a story changes in time for different values of parameters c, c_u, c_f, r and S. Of these, only the last two parameters change from one submission to another. Therefore, we fix values of the parameters $c = 0.3$, $c_u = 0.3$ and $c_f = 0.3$ and study the effect r and S have on the number of votes the story receives. We also fix the rate at which visitors visit Digg at $N = 10$ users per minute. The actual visiting rate may be vastly different, but we can always adjust the other parameters accordingly. We set the promotion threshold $h = 40$.

Our first observation is that introducing social recommendation via the Friends interface allows stories with smaller r to be promoted to the front page. In fact, we can obtain an analytic solution for the maximum number of votes a story can receive on the upcoming stories queue without the social filtering effect being present. We set $v_f = v_s = v_m = 0$ and convert Equation 7 to a differential form by taking $\Delta t \to 0$:

$$\frac{dm}{dt} = rcc_u^{k_u t} N \tag{8}$$

The solution of the above equation is $m(T) = rcN(c_u^{k_u T} - 1)/(k_u \log c_u) + 1$. Since $c_u < 1$, the exponential term will vanish for large times and leave us with $m(T \to \infty) = -rcN/(k_u \log c_u) + 1 \approx 42r + 1$. Hence, the maximum rating a story can receive on the upcoming pages only is 43. Since the threshold on Digg appears to be set around this value, no story can be promoted to the front page without other effects, such as users reading stories through the Friends interface. On average, the more reverse friends the submitter has, the smaller the minimum interestingness required for a story he submits to be promoted to the front page. Conversely, users with few reverse friends will generally have only the very interesting stories promoted to the front page. The second observation is that the more interesting story is promoted faster than a less interesting story.

Next, we consider the second modality of the Friends interface which allows users to see the stories their friends voted on. This is the situation described

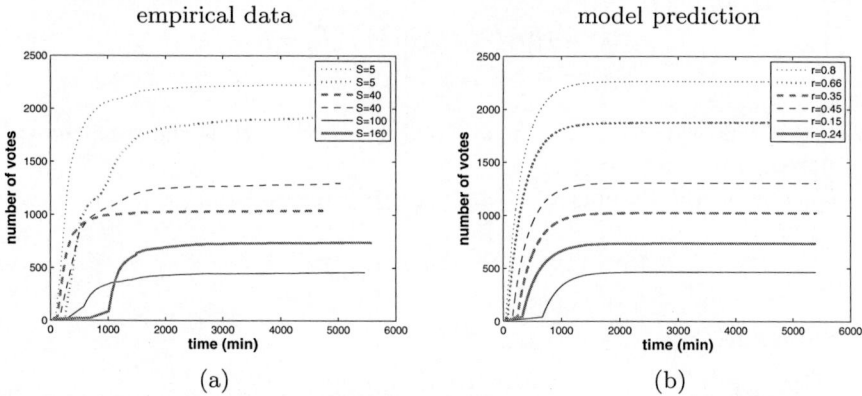

empirical data   model prediction

(a)   (b)

Fig. 4. (a) Evolution of the number of votes received by six stories from the May data set. The number of reverse friends the story submitter has is given by S. (b) Predictions the model makes for the same values of S as (a).

by the model Equation 7. Figure 4(a) shows the evolution of the number of votes received by six real stories from our data set. S denotes the number of reverse friends the story's submitter had at the time of submission. Figure 4(b) shows solutions to Equation 7 for the same values of S and values of r chosen to produce the best fit to the data. Overall there is qualitative agreement between the data and the model, indicating that the basic features of the Digg user interface we considered are enough to explain the patterns of collaborative rating. The only significant difference between the data and the model is visible in the bottom two lines, for stories submitted by users with $S = 100$ and $S = 160$. The difference between data and the model is not surprising, given the number of approximations made in the course of constructing the model (see Section 5). For example, we assumed that the combined social network of voters grows at the same rate for all stories. This obviously cannot be true. If the combined social network grew at a slower than assumed rate for the story posted by the user with $S = 160$, then this would explain the delay in being promoted to the front page. Another effect not currently considered is that a story may have a different interestingness value for users within the submitter's social network than to the general Digg audience. The model can be extended to include inhomogeneous r.

3.3   Modeling as a Design Tool

Designing a collaborative rating system like Digg, which exploits the emergent behavior of many independent evaluators, is exceedingly difficult. The choices made in the user interface can have a dramatic impact on the behavior of the collaborative rating system and on user experience. For example, should the users see the stories their friends voted on or the week's or the month's most popular stories? The designer has to consider also the tradeoffs between story timeliness and interestingness, and frequency at which the stories are promoted.

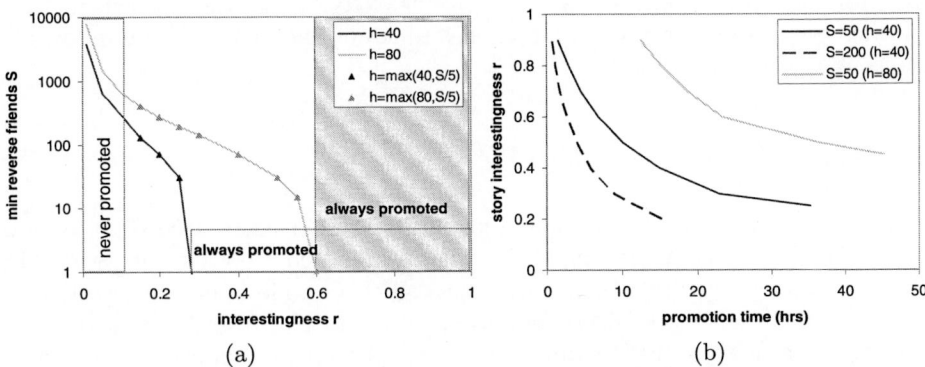

Fig. 5. (a) Minimum number of reverse friends S the submitter must have for the story he submits to be promoted to the front page for different promotion mechanisms: (1) constant threshold $h = 40$, (2) constant threshold $h = 80$, (3) variable threshold that depends on S ($h = max(40, S/5)$). For variable threshold, stories with $r \leq 0.1$ are never promoted to the front page, while for a constant threshold, they are promoted if submitted by a user with very large S. (b)

The promotion algorithm itself can have a dramatic impact on the behavior of the collaborative rating system. Digg's promotion algorithm (prior to November 2006) alienated some users by making them feel that a cabal of top users controlled the front page. Changes to the promotion algorithm appear to have alleviated some of these concerns [5] (while perhaps creating new ones). Unfortunately, there are few tools, short of running the system, that allow developers to explore the various choices of the promotion algorithm.

We believe that mathematical modeling and analysis can be a valuable tool for exploring the design space of collaborative rating systems, despite the limitations described in Section 5. Analysis can be used to compare the effects of different promotion algorithms *before* they are implemented. Figure 5(a) plots the minimum number of reverse friends, S, the submitter must have for a story he submits to be promoted to the front page for different promotion algorithms. When promotion threshold h is constant, e.g., a story needs to accumulate $h = 40$ (or $h = 80$) number of votes before being promoted, even an uninteresting story with $r \leq 0.1$ will be promoted to the front page if submitted by a well-connected user. Interesting stories ($r \geq 0.3$ for $h = 40$ and $r \geq 0.6$ for $h = 80$) will always be promoted, even if posted by unknown users. To prevent uninteresting stories from being promoted to the front page, Digg could use a variable promotion threshold that takes submitter's social network size into account. Setting $h = max(40, S/5)$ (or $h = max(80, S/5)$) prevents stories with $r \leq 0.1$ from reaching the front page.

The site designer can also investigate the effect promotion algorithm has on story timeliness. Figure 5(b) shows the time it takes a story with certain r to be promoted to the front page depending on S and the promotion threshold. The more reverse friends the submitter has, the faster the story is promoted, but the promotion algorithm has an even bigger effect. This is how mathematical analysis can help a designer evaluate the trade-offs she makes in the process of choosing the promotion algorithm.

4   Dynamics of User Rank

From its inception until February 2007, Digg ranked users according to how successful they were in getting their stories promoted to the front page. The more front page stories a user had, the higher was his standing (*rank* = 1 being the highest). If two users had an equal number of front page stories, the one who was more active (commented and voted on more stories) had higher rank. The Top Users list was publicly available and offered prestige to those who made it into the top tier. In fact, it is widely believed that improving ones rank, or standing within the community, motivated many Digg users to devote significant portions of their time to submitting, commenting on and reading stories. Top users garnered recognition as other users combed the Top Users list and made them friends. They came to be seen as influential trend setters whose opinions and votes were very valuable [15]. In fact, top users became a target of marketers, who tried to pay them to promote their products and services on Digg by submitting or voting on content created by marketers. In an attempt to thwart this practice, in February 2007 Digg discontinued making the Top Users list publicly available.

We are interested in studying the dynamics of user rank within the Digg community. For our study we collected data about the top 1,000 ranked Digg users weekly from May 2006 to February 2007. For each user we extracted user's rank, the number of stories the user submitted, commented and voted on, the

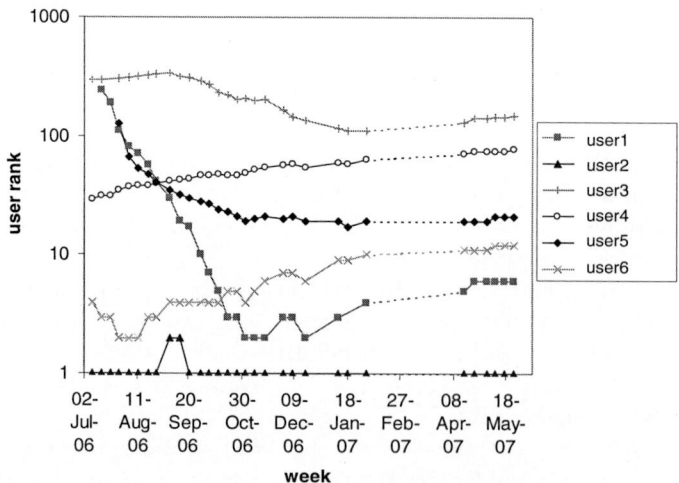

Fig. 6. Evolution of user rank for six users

number of stories that were promoted to the front page, and the number of user's friends and reverse friends ("people who have befriended the user"). Figure 6 shows the change in rank of six different users from the data set. The top ranked user (*user*2) managed to hold on to that position for most of time, but *user*6, who was ranked second at the beginning of the observation period saw his rank slip to 10. Some users, such as *user*1 and *user*5, came in with low rank but managed to reach the top tier of users by week 20. Others (*user*4 and *user*6) saw their rank stagnate.

4.1 Mathematical Model

We are interested in creating a model that can predict how a user's rank will change in time based on the user's activity level. The model also describes the evolutions of the user's personal social network, or the number of reverse friends. In addition to its explanatory power, the model can be used to detect anomalies, for example, cases when a user's rank, or social network, changes faster than expected due to collusion with other users or other attempts to game the community. Because we do not know the exact formula Digg uses to compute

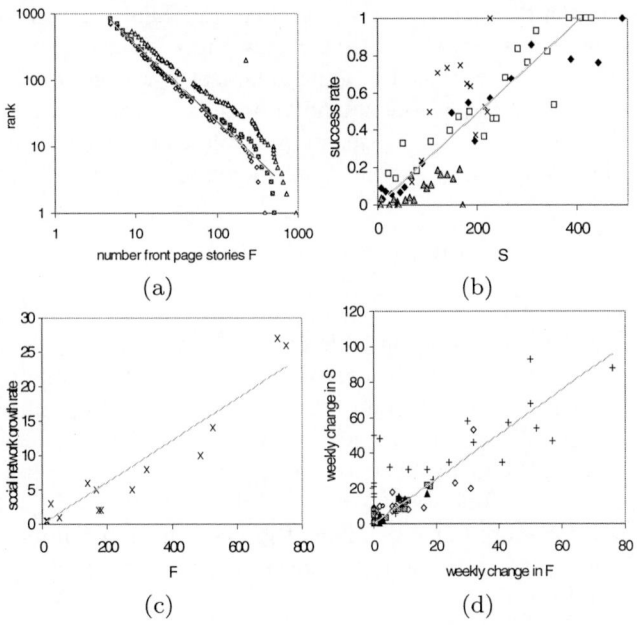

Fig. 7. Parameter estimation from data. (a) Users' rank vs number of their stories that have been promoted to the front page. (b)Different users' success rates at getting their stories promoted to the front page vs the number of reverse friends they have. In all plot, solid lines represent fit to the data. (c) Temporal growth rate of the number of user's reverse friends as a function of user rank for the weeks when no new stories were submitted by these users. (d) Weekly change in the size of the social network vs newly promoted front page stories.

rank, we will use F, the number of user's front page stories, as a proxy for rank. Figure 7(a) plots user's rank vs the number of front page stories for three randomly chosen users. The data is explained well by a power law with exponent -1: i.e., $rank \propto 1/F$.

The number of stories promoted to the front page clearly depends on the number of stories a user submits, with the proportionality factor based on the user's success rate. A user's success rate is simply the fraction of the newly submitted stories that are promoted to the front page. As we showed above, a user's success rate is linearly correlated with the number of reverse friends he has — what we call social network size S. If M is the rate of new submissions made over a period of time Δt =week, then the change in the number of new front page stories is

$$\Delta F(t) = cS(t)M\Delta t \tag{9}$$

To estimate c, we plot user's success rate vs S for several different users, as shown in Figure 7(b). Although there is scatter, a line with slope $c = 0.002$ appears to explain most of the trend in the data.

A given user's social network S is itself a dynamic variable, whose growth depends on the rate other users discover him and add him as a friend. The two major factors that influence a user's visibility and hence growth of his social network are (i) his new submissions that are promoted to the front page and (ii) his position on the Top Users list. In addition, a user is visible through the stories he submits to the upcoming stories queue and through the comments he makes. We believe that these effects play a secondary role to the two mentioned above. The change in the size of a user's social network can be expressed mathematically in the following form:

$$\Delta S(t) = g(F)\Delta t + b\Delta F(t) \tag{10}$$

In order to measure $g(F)$, how a user's rank affects the growth of his social network, we identified weeks during which some users made no new submissions, and therefore, had no new stories appear on the front page. In all cases, however, these users' social networks continued to grow. Figure 7(c) plots the weekly growth rate of S vs F. There is an upward trend indicating that the higher the user's rank (larger F) the faster his network grows. The grey line in Figure 7(c) is a linear fit to the data of functional form $g(F) = aF$ with $a = 0.03$. Figure 10(d) shows how newly promoted stories affect the growth in the number of reverse friends for several users. Although there is variance, we take $b = 1.0$ from the linear fit to the data.

4.2   Solutions

Figure 8 shows how the personal social network (number of reverse friends) and the number of front page stories submitted by six users from our data set change in time. The users are the same ones whose rank is shown in Figure 6. The plots on the right show solutions to Equation 9 and Equation 10 for each user. The

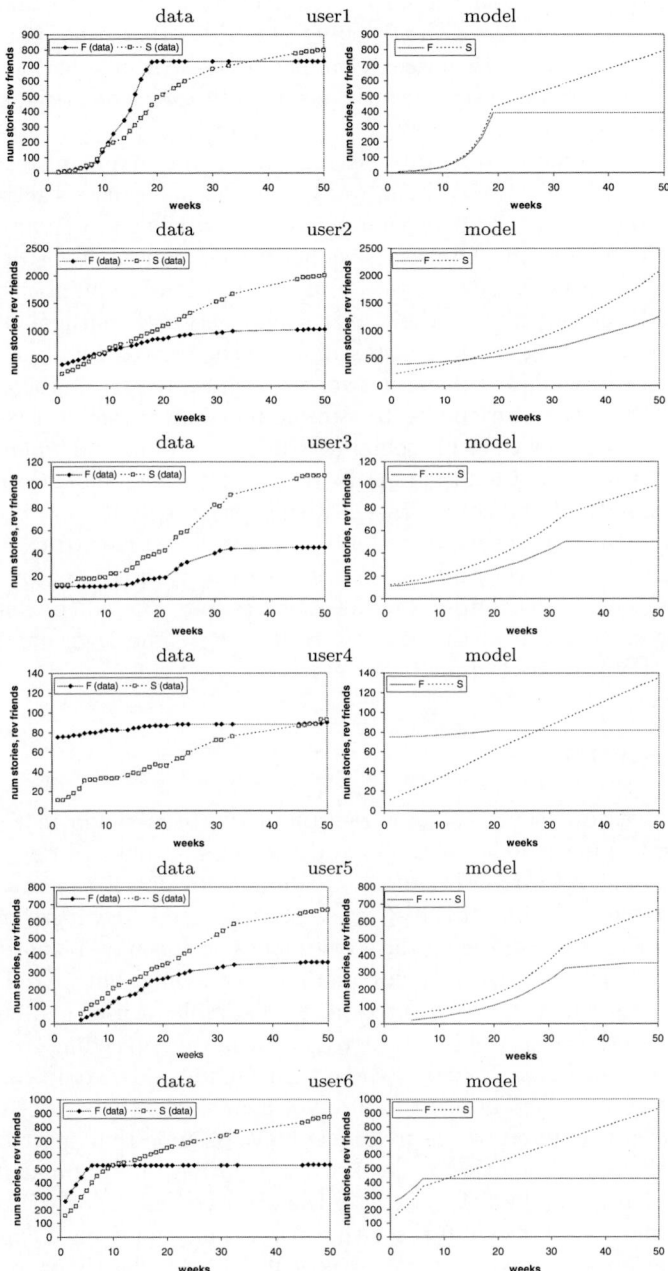

Fig. 8. Change over the course of 50 weeks in the number of front page stories and the number of user's reverse friends. The six users are the same ones whose rank is shown in Figure 6. The plots on the left show data extracted from Digg. The plots on the right show solutions to the rank dynamics model for each user.

equations were solved under the initial conditions that F and S take the values they have at the beginning of the tracking period for that user. The submission rate M was fixed at its average weekly value over the tracking period. The actual submission rate fluctuates significantly for a given user from week to week. All other parameters were kept the same for all users.

Solutions to the model qualitatively reproduce the important features of the evolution of user's rank and social network. Two factors — user's activity via new submissions and the size of his social network — appear to explain the change in user's rank. As long as the user stays active and contributes stories to Digg, as exemplified by $user2$, $user3$, $user4$ and $user5$, both the number of promoted stories (rank) and the size of the user's social network continue to grow. If a user stops contributing stories, $user1$ and $user6$, his rank will stagnate as F remains constant, while his social network continues to grow, albeit at a slower rate. Although a user can choose to submit more or fewer stories to Digg, he cannot control the growth of his social network, e.g ., how and when other users choose to make him a friend.[5] This helps promote independence of opinions, a key requirement of the collaborative rating process, and raise the quality of ratings. It appears, however, that the Top Users list serves to cement the top tier position of the highest ranked users, since they continue to grow their social networks, which in turn improves their success rate. It will be interesting to observe how elimination of the Top Users list alters the Digg community and the quality of stories that appear on the front page.

5   Limitations

A number of assumptions and abstractions have been made in the course of constructing the mathematical model and choosing its parameters. Some of our assumptions affect the structure of the model. For example, the only terms that contribute to the visibility of the story come from users viewing the front page, upcoming stories queue or seeing the stories one's friends have recently submitted or voted on. There are other browsing modalities on Digg that we did not include in the model. In the Technology section, for example, a user can choose to see only the stories that received the most votes during the preceding 24 hours ("Top 24 Hours") or in the past 7, 30 or 365 days. In the model, we only considered the default "Newly popular" browsing option, which shows the stories in the order they have been promoted to the front page. We assume that most users choose this option. If data shows that other browsing options are popular, these terms can be included in the model to explain the observed behavior. Likewise, in the Friends interface, a user can also see the stories his friends have commented on or that are still in the upcoming queue, as well as the stories they have submitted or voted on. We chose to include only the latter two options from the Friends interface in our model.

---

[5] We suspect that a user is able to influence the growth of his social network through the implicit social etiquette of reciprocating friend requests, but we have not yet been able to prove this conjecture.

In addition to the model structure, we made a number of assumptions about the form of the terms and the parameters. The first model describes the dynamics of votes an *average* story receives. In other words, it does not describe how the rating of a specific story changes in time, but the votes on many similar stories averaged together. Another point to keep in mind is that although there must exist a large variance in Digg user behavior, we chose to represent these behaviors by single valued parameters, not distributions. Thus, we assume a constant rate users visit Digg, characterized by the parameter N in the model. We also assume that a story's interestingness is the same for all users. In the model for rank dynamics, all parameters were characterized by single value — taken to be the mean or characteristic value of the distribution of user behavior. In future work we intend to explore how using distributions of parameter values to describe the variance of user behavior affects the dynamics of collaborative rating.

The assumptions we make help keep the models tractable, although a question remains whether any important factors have been abstracted away so as to invalidate the results of the model. We claim that the simple models we present in the paper do include the most salient features of the Digg users' behavior. We showed that the models qualitatively explain some features of the observed collective voting patterns. If we need to quantitatively reproduce experimental data, or see a significant disagreement between the data and predictions of the model, we will need to include all browsing modalities and variance in user behavior. We plan to address these issues in future research.

6   Previous Research

Many Web sites that provide information (or sell products or services) use collaborative filtering technology to suggest relevant documents (or products and services) to its users. Amazon and Netflix, for example, use collaborative filtering to recommend new books or movies to its users. Collaborative filtering-based recommendation systems [4] try to find users with similar interests by asking them to rate products and then compare ratings to find users with similar opinions. Researchers in the past have recognized that social networks present in the user base of the recommender system can be induced from the explicit and implicit declarations of user interest, and that these social networks can in turn be used to make new recommendations [3,12]. Social media sites, such as Digg, are to the best of our knowledge the first systems to allow users to explicitly construct social networks and use them for getting personalized recommendations. Unlike collaborative filtering research, the topic of this paper was not recommendation per se, but how social-network-based recommendation affects the global rating of information.

Social navigation, a concept closely linked to CF, helps users evaluate the quality of information by exposing information about the choices made by other users "through information traces left by previous users for current users" [2]. Exposing information about the choices made by others has been has been shown [14] to affect collective decision making and lead to a large variance in popularity of similar quality items. Unlike the present work, these research projects took into account only global information about the preferences of others (similarly to the best seller lists and Top Ten albums). We believe that exposing local information about the choices of others within your community can lead to more effective collective decision making.

Wu and Huberman [16] have recently studied the dynamics of collective attention on Digg. They proposed a simple stochastic model, parametrized by a single quantity that characterizes the rate of decay of interest in a news article. They collected data about the evolution of diggs received by front page stories over a period of one month, and showed that the distribution of votes can be described by the model. They found that interest in a story peaks when the story first hits the front page, and then decays with time, with a half-life of about a day, corresponding to the average length of time the story spends on the first front page. The problem studied by Wu and Huberman is complementary to ours, as they studied dynamics of stories *after* they hit the front page. The authors did not identify a mechanism for the spread of interest. We, on the other hand, propose social networks as a mechanism for spreading stories' visibility and model evolution of diggs both before and after the stories hit the front page. The novelty parameter in their model seems to be related to a combination of visibility and interestingness parameters in our model, and their model should be viewed as an alternative.

This paper borrows techniques from mathematical analysis of collective behavior of multi-agent systems. Our earlier work proposed a formal framework for creating mathematical models of collective behavior in groups of multi-agent systems [8]. This framework was successfully applied to study collective behavior in groups of robots [10,6]. Although the behavior of humans is, in general, far more complex than the behavior of robots, within the context of a collaborative rating system, Digg users show simple behaviors that can be analyzed mathematically. By comparing results of analysis with real world data extracted from Digg, we showed that mathematical modeling is a feasible approach to study collective behavior of online users.

7   Conclusion

The new social media sites offer a glimpse into the future of the Web, where, rather than passively consuming information, users will actively participate in creating, evaluating, and disseminating information. One novel feature of these sites is that they allow users to create personal social networks which can then be used to get new recommendations for content or documents. Another novel feature is the collaborative evaluation of content, either explicitly through voting or implicitly through user activity.

We studied collaborative rating of content on Digg, a social news aggregator. We created a mathematical model of the dynamics of collective voting and found that solutions of the model qualitatively agreed with the evolution of votes received by actual stories on Digg. We also studied how user's rank, which measures the influence of the user within the community, changes in time as the user submits new stories and grows his social network. Again we found qualitative agreement between data and model predictions. Both models were exceedingly simple. The model of the dynamics of ratings had two adjustable parameters, the rank dynamics model had none: all other model parameters were estimated from data. The agreement between models' predictions and Digg data shows that the models captured the salient features of Digg user behavior.

Besides offering a qualitative explanation of user behavior, mathematical modeling can be used as a tool to explore the design space of user interfaces. The design of complex systems such as Digg that exploit emergent behavior of large numbers of users is notoriously difficult, and mathematical modeling can help to explore the design space. It can help designers investigate global consequences of different story promotion algorithms before they are implemented. Should the promotion algorithm depend on a constant threshold or should the threshold be different for every story? Should it take into account the story's timeliness or the popularity of the submitter, etc.?

Acknowledgements. This work is based on an earlier work: "Dynamics of Collaborative Document Rating Systems", in Proceedings of the 9th WebKDD and 1st SNA-KDD 2007 workshop on Web mining and social network analysis, ACM, 2007. http://doi.acm.org/10.1145/1348549.1348555.

This research is based on work supported in part by the National Science Foundation under Award Nos. IIS-0535182 and IIS-0413321. We are grateful to Fetch Technologies for providing wrapper building and execution tools, and to Dipsy Kapoor for processing data.

References

1. Adar, E., Zhang, L., Adamic, L.A., Lukose, R.M.: Implicit structure and the dynamics of blogspace. In: Workshop on the Weblogging Ecosystem, 13th International World Wide Web Conference (2004)
2. Dieberger, A., Dourish, P., Höök, K., Resnick, P., Wexelblat, A.: Social navigation: techniques for building more usable systems. Interactions 7(6), 36–45 (2000)
3. Kautz, H., Selman, B., Shah, M.: Referralweb: Combining social networks and collaborative filtering. Communications of the ACM 4(3), 63–65 (1997)
4. Konstan, J.A., Miller, B.N., Maltz, D., Herlocker, J.L., Gordon, L.R., Riedl, J.: GroupLens: Applying collaborative filtering to Usenet news. Communications of the ACM 40(3), 77–87 (1997)
5. Lerman, K.: Social information processing in social news aggregation. IEEE Internet Computing: special issue on Social Search 11(6), 16–28 (2007)
6. Lerman, K., Jones, C.V., Galstyan, A., Matarić, M.J.: Analysis of dynamic task allocation in multi-robot systems. International Journal of Robotics Research 25(3), 225–242 (2006)
7. Lerman, K., Jones, L.: Social browsing on flickr. In: Proc. of International Conference on Weblogs and Social Media (ICWSM 2007) (2007)
8. Lerman, K., Martinoli, A., Galstyan, A.: A review of probabilistic macroscopic models for swarm robotic systems. In: Şahin, E., Spears, W.M. (eds.) Swarm Robotics 2004. LNCS, vol. 3342, pp. 143–152. Springer, Heidelberg (2005)

9. Lerman, K., Shehory, O.: Coalition Formation for Large-Scale Electronic Markets. In: Proceedings of the International Conference on Multi-Agent Systems (ICMAS 2000), Boston, MA, pp. 167–174 (2000)
10. Martinoli, A., Easton, K., Agassounon, W.: Modeling of swarm robotic systems: A case study in collaborative distributed manipulation. Int. Journal of Robotics Research 23(4), 415–436 (2004)
11. Mika, P.: Ontologies are us: A unified model of social networks and semantics. In: Gil, Y., Motta, E., Benjamins, V.R., Musen, M.A. (eds.) ISWC 2005. LNCS, vol. 3729, pp. 522–536. Springer, Heidelberg (2005)
12. Perugini, S., André Gonalves, M., Fox, E.A.: Recommender systems research: A connection-centric survey. Journal of Intelligent Information Systems 23(2), 107–143 (2004)
13. Rose, K.: Talk presented at the Web2.0 Conference (November 10, 2006)
14. Salganik, M.J., Dodds, P.S., Watts, D.J.: Experimental study of inequality and unpredictability in an artificial cultural market. Science 311, 854 (2006)
15. Warren, J., Jurgensen, J.: The wizards of buzz. Wall Street Journal online (February 2007)
16. Wu, F., Huberman, B.A.: Novelty and collective attention. Technical report, Information Dynamics Laboratory, HP Labs (2007)

Applying Link-Based Classification to Label Blogs*

Smriti Bhagat, Graham Cormode, and Irina Rozenbaum

Rutgers University, NJ, USA
{smbhagat,rozenbau}@cs.rutgers.edu, graham@dimacs.rutgers.edu

Abstract. In analyzing data from social and communication networks, we encounter the problem of classifying objects where there is explicit link structure amongst the objects. We study the problem of inferring the classification of all the objects from a labeled subset, using *only* link-based information between objects.

We abstract the above as a labeling problem on multigraphs with weighted edges. We present two classes of algorithms, based on local and global similarities. Then we focus on multigraphs induced by blog data, and carefully apply our general algorithms to specifically infer labels such as age, gender and location associated with the blog based only on the link-structure amongst them. We perform a comprehensive set of experiments with real, large-scale blog data sets and show that significant accuracy is possible from little or no non-link information, and our methods scale to millions of nodes and edges.

Keywords: Graph labeling, Relational learning, Social Networks.

1   Introduction

In recent years there has been rapid growth in the massive networks which store and encode information—most obviously, the emergence of the World Wide Web, but also in the social networks explicit or implicit from collections of emails, phone calls, blogs, and services such as MySpace and Facebook. It is now feasible to collect and store such data on a scale many orders of magnitude greater than initial hand-curated collections of friendship networks by sociologists.

Many fundamental problems in analyzing such data can be modeled as instances of classification. That is, we wish to attach a label to each entity in the data based on a small initially labeled subset. This is motivated by applications as varied as (web) search, marketing, expert finding, topic analysis, fraud detection, network service optimization, customer analysis and even matchmaking. The problem of "link-based classification" [10], or "relational learning" [19], has received significant study: we later survey the most relevant works. However, such studies have typically been over relatively small collections, i.e. thousands of carefully curated data items such as movies, companies, and university web pages.

In this paper, we address the challenge of labeling network data in the face of substantially larger datasets, motivated by the examples of large social networks. In particular, we study the networks induced by blogs, and the rich structure they create, although

---

H. Zhang et al. (Eds.): WebKDD/SNA-KDD 2007, LNCS 5439, pp. 97–117, 2009.
© Springer-Verlag Berlin Heidelberg 2009

our methods and observations are sufficiently general to be more widely applicable. Our choice of blogs is motivated by our earlier experiences with this data source [3]: working closely with this data reveals questions to serious analysis, such as:

- How to work across multiple networks?

Typically, entities of interest do not reside within a single network, but are spread across multiple interlinked networks. For example, blogs are hosted by multiple different providers, and link offsite to websites; in the telecoms world, calls are handled by a mixture of long distance and wireless providers. Although one can focus in on just a single network, this will often lose vital information, and lead to a disconnected view of the data. Instead, one needs to work across networks (and hence develop data collection techniques for each network type) to see a fuller picture, and recognize that different (sub)networks have different behavior.

- What information may be used to infer labels?

We focus on situations where there are explicit links between objects, and study how to use these to propagate labels. We use the fact that there is often some amount of initially labeled data, such as customer information in telecoms, or profile information for blogs. In some applications, there may be additional features, such as text in blogs. However, in many cases we can obtain no features other than graph edges: for technical, legal, or privacy reasons we often see only the links, and little additional data associated with them. In this study, we explore the power of inference based solely on link information, and conclude that for many applications this alone can give a powerful result.

- How to collect data and ensure data quality?

Data collection is a non-trivial part of applying any classification algorithm. Especially when collecting data self-reported by individuals, (e.g. from web repositories), there are issues of reliability in trustworthiness in the collected values. Additionally, one has to be aware of an inherent bias in the data due to the perspective of the observer, since some networks are easier to poll, or yield more information, than others. For example, we must choose which subset of blog hosting sites to crawl and write parsers for, and what crawling strategy to employ.

- How to scale to large data sizes?

With numbers of nodes in the hundreds of thousands, and edges in the millions, many sophisticated structural analysis techniques, such as matrix-style decompositions, fail to scale. Moreover, we also need methods which can provide relatively simple explanations of their inferences if we are to justify them to decision makers. Here, we seek to understand the power of relatively simple methods whose computational cost is near linear in the input size.

Our Contributions. Our study here is on the problem of inferring the classification labels for objects with explicit structure linking them. We can view this a problem over enormous graphs, or more correctly, multigraphs, due to the multiple object types, edge types, and multiple edges inherent in any detailed study of such networks. Our main contributions are as follows:

1. We formalize our problem of graph labeling as an instance of (semi-)supervised learning, and introduce two simple classes of algorithms, *local iterative* and *global nearest neighbor*.

2. We give a detailed description of data collection and modeling in order to apply these algorithms to blog data and to infer labels such as age, location and gender, given the challenges outlined above.

3. We show experimental results with large scale blog data that demonstrate these methods are quite accurate using link information only, and are highly scalable: for some tasks we can accurately assign labels with accuracy of 80-90% on graphs of hundreds of thousands of nodes in a matter of tens of seconds.

Our chosen methods succeed due to inherent structure in the blog network: people tend to link to others with similar demographics, or alike people link to similar objects. It remains a challenge to extend them to situations where such behavior is less prevalent, and understand fully when they are and are not applicable.

Outline. In what follows, we will describe the abstract problem of labeling (multi) graphs and our two classes of solutions in Section 2. In Section 3, we will describe the study case of blogs and describe how our algorithms are applicable for inferring labels on blogs. In Section 4, we present extensive experiments with large scale blog data to infer age/location/gender of blogs. Related work is discussed in Section 5 and concluding remarks are in Section 6.

2 Graph Labeling

2.1 Problem Formulation

We define the problems of graph and multigraph labeling. Given a graph, with a subset of nodes labeled, our goal is to infer the labels on the remaining nodes. Formally,

Definition 1. *Let G be a partially labeled directed graph $G = (V, E, M)$, where V is the set of vertices or nodes and E is the set of edges as usual. Let $L = \{\ell_1, \ell_2, ..., \ell_c\}$ be the set of labels, where a label ℓ_k can take an integer or a nominal value, and $|L|$ is the number of possible labels. $M : V \to L \cup \{0\}$ is a function which gives the label for a subset of nodes $W \subset V$; for nodes $v \notin W$, $M(v) = 0$, indicating that v is initially unlabeled. The* Graph Labeling Problem *is, given the partially labeled graph G, to complete the labeling: i.e., to assign labels to nodes in $U = V \backslash W$.*

This abstract problem captures our core question: how to use the link structure of the graph and the partial labels in order to infer the remaining labels. This can be seen as an instance of Relational Learning [10], since a graph can be encoded as a (many-to-many) relation. However, we find it more useful to phrase the problem in terms of graph nodes and edges, since in our setting, the graph structure is part of the input, and may have edges that connect otherwise dissimilar nodes. This is also fundamentally a *semi-supervised* classification problem, since in our motivating applications it is rare to find any representative completely labeled graph; instead, we have one large graph with a subset of its nodes labeled. The connections to relational learning and semi-supervised classification is discussed further in Section 5.

In practice, the data that we collect is richer than this, and does not fit into the simplistic graph model. We see different kinds of nodes, relating to different entities. For example, blogs have links to webpages which are quite different from blog pages. When we have multiple types of node, and different kinds of edge connecting them, we have a *multigraph* version of the problem. An important detail is that for some node types, we have more or less information than others. This is a result of how much can be sampled or observed in the world. For example, in the telecommunications world, service providers can observe both incoming and outgoing calls for their customers, but do not see calls between customers of other providers. As a result the graph a provider sees may not contain all the outgoing/incoming edges of some of the nodes. Likewise in blog or web analysis, one may know all outgoing edges for each page, but not all the incoming edges. This gives an unavoidable bias that is due to problems of *observability* (some links are not observable) and *collectability* (it is not feasible to collect and store every last link).

Formally, we have a multigraph, $G^+ = (V^+, E^+)$, where V^+ is partitioned into p sets of nodes of different types, $V^+ = \{V_1, V_2, ..., V_p\}$, and E^+ is a (weighted) collection of sets of edges $E^+ = \{E_1, E_2, ..., E_q\}$. We can have additional features F on each node, and a function w giving the weight of each edge or set of edges. Other variations, such as features and labels on edges are possible, but we omit them to focus on the underlying problems. Some examples of multigraphs in different settings are:

- **Telecommunications:** The nodes of the multigraph may represent distinct phone numbers, and the edges represent telephone calls made between two phone numbers (clearly directional). One node type may represent 1-800 numbers that can only receive calls, while the other nodes are consumer accounts. There can be multiple edges between nodes and multiple kinds of edges (long distance calls, local calls and toll free calls). A suitable label in this example is to classify the numbers as business/non-business. Typically telephone companies have a business directory to populate labels on a subset of nodes, and in some cases, use human evaluation to label some nodes too.
- **IP networks:** In the IP network setting, a node could represent a distinct IP address, a segment of IP addresses or an ISP. An edge between two nodes may signify any kind of IP traffic detected between the two nodes, traffic belonging to a certain application or protocol, certain types of messages, etc. A suitable label in this case is based on the network node's function as a server or a client. Typically Internet Service Providers have a list of known or suspected servers which is the initial set of labels from which we need to infer the classification of server/client for remaining nodes.
- **Web:** The World Wide Web can be represented by a multigraph, where nodes are webpages that can be further categorized by ownership, functionality or topic [5], and an edge between two nodes signifying an HTML link from one web page to another. Links could also be categorized: e.g., an edge from a site to a commercial company website could signify presence of the company's advertisement on the website. In this setting, suitable node labels could be based on the site being public or commercial, or the site's function (portal, news, encyclopedia, etc). The class of some nodes are known, and these can be used to label the remaining nodes. □

2.2 Algorithmic Overview

We principally study two classes of algorithms for predicting labels on multigraphs, namely, *Local Iterative* and *Global Nearest Neighbor*. These classes have their antecedents in prior work on relational learning [12] and the well-known general purpose classifiers such as Nearest Neighbors. They are designed to be relatively simple to implement and scale to very large graph data. Both the local and global approaches use the link structure and neighborhood information to infer labels. For each unlabeled node they examine a set of labeled nodes and select one of these labels as the new label. The approaches differ in the set of nodes considered for inferring labels. We call these 'classes of algorithms' as there are many feasible variations in each class which do not change the fundamental character of the algorithms.

Preliminaries. Our algorithms take as input a description of the (multi)graph and any features and labels. Since the graphs we consider may become very large in size, typically the description will be in the form of an adjacency list or similar set of edges. However, it is often convenient to give the description of the algorithms in adjacency matrix notation. Let A be an $n \times n$ adjacency matrix representing a graph $G = (V, E)$, where $a_{ij} = 1$ if $(i, j) \in E$ and 0 otherwise (more generally, a_{ij} can be a weight of an edge from i to j if any); $n = |V|$ is the number of nodes. Let $A_{(i)}$ denote the ith row of matrix A and $A^{(j)}$ denote the jth column of A, where $i, j \in [1, n]$. For a function f whose domain is $\mathrm{dom}(f)$ we use the notation $\mathrm{diag}(f)$ as shorthand for the $\mathrm{dom}(f) \times \mathrm{dom}(f)$ matrix such that $\mathrm{diag}(f)_{ii} = f(i)$, and is zero elsewhere.

The neighborhood of each i node, defined by the immediately adjacent nodes, is encoded as a *feature vector*, $B_{(i)}$, based on the link structure of i (in general, the feature vector could also include other features of the node, but here we focus on a core link-based classification problem). Typically, this feature vector is initialized to represent the frequency of the labels on the nodes in its neighborhood. From these vectors, we create an $n \times c$ feature matrix B. Given a function f mapping from n to c, let $\chi(f)$ denote the characteristic matrix of f, i.e. $\chi(f)_{il} = 1$ iff $f(i) = l$. We can write $B = A\chi(M)$, where M is the initial labeling. In such graph labeling problems we potentially *change* the feature vectors as we make inferences and label nodes—when the neighbor of node i is labeled, the feature vector of i changes to reflect this new label—to enable the propagation of labels to all nodes.

We consider only directed graphs, since the natural extension of these algorithms to the undirected case is equivalent to the directed case where each directed edge is present in the reverse direction. However, because of this directionality, nodes with no incoming edges have no neighborhood information for us to predict from (they have an empty feature vector); hence such nodes may remain unlabeled, or can be labeled with most likely label from the prior distribution. In some networks, link reciprocity (a bi-directional link) is an important indicator of a stronger connection; this can be captured giving an increased weight to reciprocal links.

2.3 Local Iterative Methods

With only a graph and some initial labels, there is limited information available to help us propagate the labeling. We therefore make some assumptions about how nodes attain

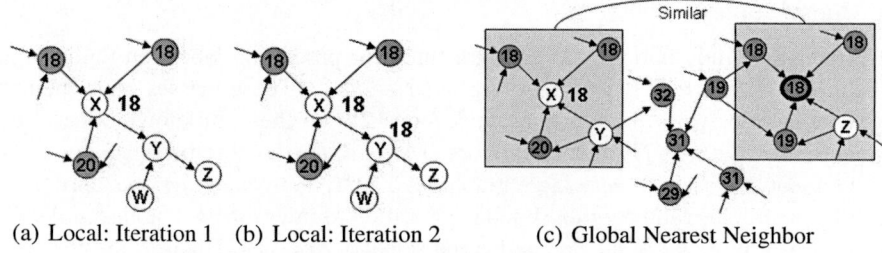

(a) Local: Iteration 1   (b) Local: Iteration 2   (c) Global Nearest Neighbor

Fig. 1. Graph Labeling Examples

their labels. Our local methods assume the existence of *homophily*: that links are formed between similar entities so that the label of a node is a function of the labels of nearby nodes [13]. The algorithms succeed if the following hypothesis holds:

> Nodes link to other nodes with similar labels.

Thus, we view each incoming edge in a directed graph as representing a "vote" by an adjacent node. Let ℓ_k be the label on node u. An edge from node u to node v implies node u is a vote for the label ℓ_k for node v. For each unlabeled node, the votes from its neighbors are combined to derive its label by a function $\text{voting}(B_{(i)})$. A single application of this procedure results in labeling some unlabeled nodes, but not every node receives a label instantly. Therefore, we recalculate feature vectors for nodes in the neighborhood of a newly labeled node, and iterate. When performed iteratively, recalculating feature vectors each step, this leads to propagation of labels. There are many variations, based on the voting function used to assign the label based on the feature vector, the neighborhood used to generate the feature vector, and so on. Intuitively, the simplest voting scheme is "plurality voting" which chooses the most frequently voted label by the immediately adjacent nodes of an unlabeled node (as in [12]). We can also consider other functions, such as taking the median or average label drawn from an ordered domain, and so on.

In each iteration, we visit each node $i \in U$, and determine a label for it based on its current neighborhood, rather than fixing the label for an unlabeled node once and for all. A node may receive different labels in different steps, with different levels of confidence in each based on the round in which the label is assigned. Clearly, a neighbor of i which is initially labeled is more reliable than one which gets a label in a late iteration.

Formal Definition. Let A be the adjacency matrix representation of the graph. At each iteration, we compute a new labeling function M^t such that for every (unlabeled) node $i \in U$, the label determined based on voting by its neighbors is assigned to $M(i)$. To label the nodes, at iteration t, M^t is defined by:

$$M^t(i) \leftarrow \text{voting}(B^t_{(i)})$$

where B^t is the feature matrix for the t-th iteration, defined by:

Definition 2. *Let $M^t : V \rightarrow L \cup \{0\}$ denote the labeling function on the t-th iteration (we insist that $M^t(i) = M(i)$ for $i \in W$). Let $\text{conf}^t : V \rightarrow \mathbb{R}$ be a function from nodes*

Algorithm 2.1: LOCALITERATIVE(E, M, s)

$$B^0 \leftarrow 0; \quad M^0 \leftarrow M$$

for $t \leftarrow 1$ **to** s

$\quad$ **do** $\begin{cases} \textbf{for } i \in U \\ \quad \textbf{do} \begin{cases} \textbf{for } (i,j) \in E \\ \quad \textbf{do} \begin{cases} k \leftarrow M^{t-1}(j) \\ \textbf{if } k \neq 0 \\ \quad \textbf{then } B_{ik}^t \leftarrow B_{ik}^t + 1 \end{cases} \end{cases} \\ \textbf{for } j \in V \\ \quad \textbf{do} \begin{cases} \textbf{if } j \in U \\ \quad \textbf{then } M^t(j) \leftarrow \text{voting}(B_{(j)}^t) \\ \quad \textbf{else } M^t(j) \leftarrow M(j) \end{cases} \\ B^{t+1} \leftarrow B^t \end{cases}$

Algorithm 2.2: GLOBAL1NN(E, M)

$$B_{n \times c} \leftarrow 0; \quad S_{n \times N} \leftarrow 0$$

for $(i,j) \in E$

$\quad$ **do** $\begin{cases} k \leftarrow M(j) \\ \textbf{if } k \neq 0 \\ \quad \textbf{then } B_{ik} \leftarrow B_{ik} + 1 \end{cases}$

for $i \in U$

$\quad$ **do** $\begin{cases} \textbf{for } j \in W \\ \quad \textbf{do } S_{ij} \leftarrow sim(B_{(i)}, B_{(j)}) \\ k \leftarrow \arg\max(S_{(i)}) \\ M(i) \leftarrow M(k) \end{cases}$

(a) Local Iterative Algorithm

(b) Global Nearest Neighbor Algorithm

Fig. 2. Local and Global Algorithms

denoting the relative confidence of the algorithm in its labeling at the t-th iteration. We set $M^0 = M$ and $\text{conf}^0(i) = 1$ for all $i \in W$, zero otherwise. Lastly, let decay $: \mathbb{N} \rightarrow \mathbb{R}$ *be a function which returns a weighting for all labels assigned a iterations ago. We define the* iterative feature vector at iteration t, B^t *as*

$$B^t = A \sum_{t'=0}^{t-1} \text{decay}(t - t') \chi(M^{t'}) \, \text{diag}(\text{conf}^{t'}) \qquad (*)$$

We consider a simple setting of these functions: our instantiation assigns equal confidence to every labeling (i.e. $\text{conf}^t(i) = 1$ for all i, t), and equal weighting for all label values (i.e. $\text{decay}(x) = 1$ for all x), which favors labels assigned earlier in the process. This makes B^t convenient to compute: (*) simplifies to $B^t = B^{t-1} + AM^{t-1}$, and so can be computed in place without keeping a history of prior values of $B^{t'}$. Note that the absolute values of entries in B^t are unimportant, just the relative values, due to our application of voting via $\arg\max$ (thus one can normalize and view B^t as a distribution on labels, and the voting take the maximum likelihood label). We set a bound on the number of iterations performed before terminating as s. Algorithm 2.1 shows pseudocode for the local iterative method. In summary (proofs omitted for brevity):

Lemma 1. *Each iteration of the local iterative algorithm runs in $O(|V| + |E|)$ time.*

Example. Figures 1(a) and 1(b) show the working of the local iterative approach on a small example, with $\arg\max$ as the voting function. Nodes X, Y, Z and W are unlabeled. The feature of vector of node X records a frequency of 2 for label '18', and 1 for label '19'. So in the first iteration, node X is labeled '18' based on the votes by its neighbors. In the following iterations, labels are propagated to nodes Y and Z. Since node W does not have any incoming edges, it remains unlabeled. □

Iterations. The number of iterations should be chosen large enough so that each node has the chance to get labeled. This is bounded by the diameter of the (directed) graph and the graphs we consider are typically "small world", and so have a relatively small diameter—often, $O(\log |V|)$. Additional iterations increase our confidence in the labeling, and allow the labeling to stabilize. Although is possible to create adversarial examples where the iterative process never stabilizes (because the label of a node cycles through several possible labelings), such extreme cases do not occur in our experiments.

Multigraph Case. When there are multiple node and edge types these issues arise:

• **Pseudo-labels.** In multigraphs, we may have multiple classes of nodes, where the label only applies to certain of them. In our telecoms example, it is not meaningful to classify the calling patterns of certain numbers (such as 1-800 numbers which are for incoming calls only). Instead of omitting such nodes, we use the notion of *pseudo-labels*: allocating labels to nodes by the iterative method even if these labels are not wholly meaningful, as a means to the end of ensuring the nodes of interest do receive meaningful labels. This generalization turns out to be very important in propagating labels to otherwise isolated nodes.

• **Edge weights.** With different types of edge, and different numbers of edges between nodes, it is natural to introduce edge weights, and modify the feature vectors by using these weights to scale the votes. These details are mostly straightforward.

• **Additional Features.** In some cases we have additional features F attached to certain nodes or edges. It is not completely obvious how to cleanly extend the iterative approach to incorporate these. One direction is to use an appropriate distance function to measure the similarity between the features of pairs of nodes connected by an edge, and re-weight the voting based on the similarity: the votes of more similar nodes count more than others. However, the specification of the distance and weighting schemes is non-trivial, and is left for future work.

2.4 Global Nearest Neighbor

The global nearest neighbor family of methods uses a different notion of proximity to the local algorithm to find labels. By analogy with the traditional k-Nearest Neighbor classifier, we consider the set of labeled nodes and take the labels of the k-best matches. The matching is based on the similarity of the feature vectors, i.e. the labels of the neighboring nodes. The underlying hypothesis is one of *co-citation regularity* [10], i.e.:

> Nodes with similar neighborhoods have similar labels.

Example. Figure 1(c) gives an example of labeling by the global 1-nearest neighbor approach. Of all labeled nodes, the neighborhood of nodes X is most similar to that of the highlighted node with label 18. The algorithm assigns the label 18 to node X. A similar nearest neighbor search is repeated for all unlabeled nodes.   □

Formal Description. As before, consider the adjacency matrix A and an initial labeling function M, which define a feature matrix B. Let $S_{n \times n}$ be a similarity matrix. For each unlabeled node i, compute the similarity coefficient S_{ij} between $B_{(i)}$ and $B_{(j)}$, for each labeled node j. Node i is assigned the most frequent label of the k nodes with the highest similarity coefficients i.e. the label of node $\arg\max(S_{(i)})$.

Choice of Similarity Function. Given two vectors x and y, there are many possible choices, such as the L_p distances: Euclidean distance, $\|x-y\|_2$, and Manhattan distance, $\|x-y\|_1$. We employ Pearson's correlation coefficient,

$$C(x,y) = \frac{nx \cdot y - \|x\|_1\|y\|_1}{\sqrt{n\|x\|_2^2 - \|x\|_1^2}\sqrt{n\|y\|_2^2 - \|y\|_1^2}}.$$

Intuitively, the correlation coefficient is preferred over Euclidean distance when the shape of the vectors being compared is more important than the magnitude.

In the multigraph case, we can naturally take into account different nodes and edges (V^+, E^+) and features F by keeping the algorithm fixed and generalizing the similarity function. For set valued features, we can compare sets X and Y using measures such as Jaccard $(J(X,Y) = \frac{|X \cap Y|}{|X \cup Y|})$. The similarity function can combine the similarities of each feature. For example, we later utilize a weighted combination of Jaccard coefficient (for features represented as sets) and correlation coefficient (for vector features). Algorithm 2.2 shows pseudocode for the global nearest neighbor method. In summary:

Lemma 2. *The running time for global nearest neighbor is $O(|U||W||L| + |E|)$.*

Approximate Nearest Neighbors. The above lemma assumes a naive exhaustive comparison of every labeled node with every unlabeled node. For appropriate similarity functions, this can be accelerated via dimensionality reduction and approximate nearest neighbors algorithms to find nodes that are approximately the closest [9].

Multi-pass Generalization. As defined above, the global nearest neighbor algorithm takes a single pass and attempts to assign a label to every unlabeled node based on the initially labeled neighborhoods. This could result in poor inferences when a node has few, if any labeled neighbors. As in the iterative case, it is possible to define a multi-pass algorithm, which bases its conclusions on labels (and confidences) defined in the previous iteration; here we focus on the single-pass version for brevity.

3 Labeling Blogs

3.1 Anatomy of a Blog

A blog is typically a web-based journal, with entries (posts) displayed in reverse chronological order. Postings are publicly viewable, and readers may provide immediate feedback by adding a comment. Websites offer blog hosting with a variety of user interfaces and features. Blogs commonly include information about the owner/author in the form of a *profile*, in addition to the blog entries themselves.

User Profile. When users opens accounts at a blog hosting site, they are asked to fill out a user profile form with age, gender, occupation, location, interests (favorite music, books, movies, etc.). In some cases, the user can also provide an email address, URL

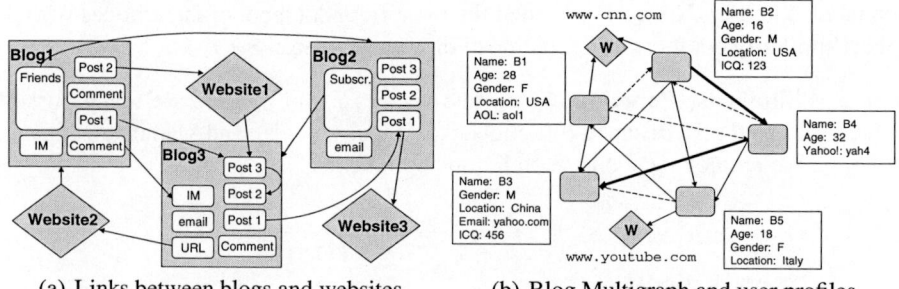

(a) Links between blogs and websites   (b) Blog Multigraph and user profiles

Fig. 3. Multigraphs extracted from blog data

of a personal website, Instant Messenger ID's, etc. Most of this information is optional. Some services only reveal some information to a set of "friends" (accounts on the same service); this list of friends may be visible to all.

Blog Entries. The blog owner posts blog entries which contain text, images, links to other websites and multimedia etc. They are typically accompanied by the date and time each entry was made. Blog postings often reference other blogs and websites (as illustrated in Figure 3(a)). Bloggers can also utilize special blog sections to display links of particular interest to them, such as "friends", "links", "subscriptions", etc.

3.2 Modeling Blogs As Graphs

There are many ways to extract a (multi)graph from a collection of blog data. We outline some of the choices in the modeling and extraction of features.

Nodes. We can encode blogs as graph nodes at several granularities: we can treat each blog posting and comment as separate nodes, or consider all postings within a single blog as a single node. There is also some subtlety here, since some blogs may have multiple authors, and single authors may contribute to multiple blogs. However, the common case is when a single author has a single blog. We choose to represent a blog by a single node. Additional nodes represent webpages connected to blogs.

Edges. We use the (web)links to define the edges in the blog graph: a directed edge in the blog graph corresponds to a link from the blog to another blog or website. We can automatically categorize these links according to the destination and where they appear within the blog pages: a link appearing in a blog entry, a link appearing in a comment posted as a response to a blog entry, a link in the "friends" category, etc. These define various sets of edges: E_F, based on explicit friend links, E_B, containing all other links to blogs, and E_W, containing links from blogs to websites.

Labels. Having defined the nodes and edges, we consider a variety of labels. In full generality, we can consider almost any label that can be attached to a subset of the nodes and propagated by our algorithms. In our study, we restrict ourselves to labels based on components of the user profile. These cover a broad set of different label types (binary, categorical, continuous), and ensure that we can use collected data to define training and test data. We consider the following labels:

• **Age.** Blog profiles typically invite the user to specify their date of birth, and a derived age is shown to viewers. But the "age" we attach to a blog can have multiple interpretations: the actual age of the blog author, the "assumed" age of the author, the age of the audience, and so on. We will evaluate our algorithms at matching given age labels that are withheld from the algorithm.

• **Gender.** Gender is another natural profile entry to attempt to propagate. Prior work has looked for text and presentation features [18] in order to predict gender; here, we aim to use link information only. Gender has multiple interpretations similar to age.

• **Location.** The (stated) location of the author can be at the granularity of continents (category with seven values) or country (over two hundred possible values).

Many other labels are possible, but we focus on these three, since they demonstrate common label types, and are available in many blog profiles for evaluation. Figure 3 illustrates possible graphs on the same set of nodes, showing different labels.

3.3 Algorithms Applied to the Blog Graph

We represent a collection of blogs and the links between them by a graph over a set of nodes V_B. Each node $v \in V_B$ corresponds to a blog user (identified by a unique user id). We examined the three label types: age, gender and location (by country or by continent). Label values are extracted from the blog owner's profile if present. In applying our algorithms, different issues arise for each label:

Age. When working with age label, our hypotheses translate to *bloggers tend to link to other bloggers of their own age* (local iterative) and *bloggers of the same age link to bloggers of similar age distributions* (global nearest neighbor). Both of these seem plausible, but distinct. The initial feature matrix, $B_{n \times 120}$, encodes the frequency of adjacent ages in years for each node. Because age is a continuous attribute, we smoothed each feature vector by convolution with the triangular kernel $[0.2, 0.4, 0.6, 0.8, 1.0, 0.8, 0.6, 0.4, 0.2]$. This improved the quality of the observed results, so all experiments shown use this kernel. Our default similarity method is the correlation coefficient.

Location. The hypotheses for the location label are that *bloggers tend to link to other bloggers in their vicinity* (local) and *bloggers in the same locale link to similar distributions of locations* (global). The former hypothesis seems more intuitively defensible. Feature vectors encode 245 countries belonging to seven continents.

Gender. For gender, the iterative assumption is *bloggers link to bloggers of their own gender* (local) or *bloggers of the same gender link to similar patterns of genders*. Neither seems particularly convincing, and indeed for gender we saw the worst experimental results. This is partly due to using the desired label as the (only) feature in the classification. As noted in Section 2.3, our methods allow other features to be included and we hypothesize that including additional features (such as the age and location, if known) could improve the learning. We defer such studies to future work.

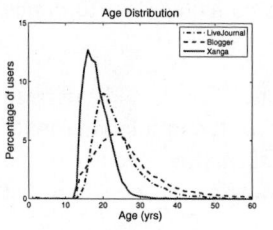
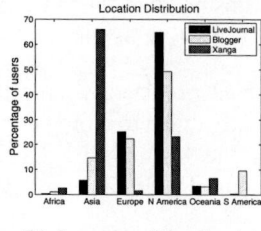
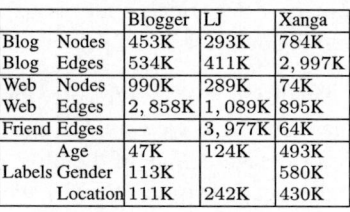

		Blogger	LJ	Xanga
Blog	Nodes	453K	293K	784K
Blog	Edges	534K	411K	2,997K
Web	Nodes	990K	289K	74K
Web	Edges	2,858K	1,089K	895K
Friend	Edges	—	3,977K	64K
	Age	47K	124K	493K
Labels	Gender	113K		580K
	Location	111K	242K	430K

(a) Age Distribution   (b) Location Distribution   (c) Summary of blog data

Fig. 4. Label and Data Collection Summary

Multigraph Edges. We tried several weighting schemes for edges reflecting the relative importance of different links (blog, friends, web), Many weightings are possible; for brevity we report only these settings: Blog-blog links only (all edges in E_B have weight 1, all other edges have weight 0); Friends only (E_F edges weight 1, others 0); Blogs and friends (E_B and E_F have weight 1); Blogs and web (E_B and E_W have weight 1).

Multigraph Nodes. In addition to nodes corresponding to blogs, we include additional nodes V_W corresponding to (non-blog) websites. These turn out to be vital in propagating labels to otherwise weakly linked subgraphs. To make use of the webpage nodes, we replicate edges to webpages in the reverse direction. All webpage nodes are initially unlabeled, and the local iterative method therefore assigns a *pseudo-label* to all of these nodes in the course of its operation. Although some labels such as Location or Age have unclear semantics when applied to webpages, it is possible to interpret them as a function of the location and age of the bloggers linking to the webpage. This can be viewed as applying co-citation regularity in the iterative model, allowing labels to be transfered from one blog to another via these intermediaries.

The Global Nearest Neighbor algorithm takes a different approach to using the nodes in V_W. Since no such nodes are initially labeled, they would play no part in the (single-pass) algorithm even if we assign them pseudo labels. Instead, we effectively treat the links to nodes in V_W as defining a set of (sparse, high dimensional) binary features. We therefore extend the similarity function between two nodes, as suggested in Section 2.4 as a weighted sum of the (set) similarity between V_W neighborhoods and (vector) similarity between V_B neighborhoods. For nodes $i \in U$ and $j \in W$ the similarity coefficient is:

$$S_{ij} = \alpha \times C(B_{(i)}, B_{(j)}) + (1 - \alpha) \times J(V_{W(i)}, V_{W(j)})$$

for some $0 \le \alpha \le 1$, where $B_{(i)}$ is the feature vector of the node i, and $V_{W(i)}$ is the set of web nodes linked to the blog node i.

4   Experiments

4.1   Experimental Setup

Data Collection. In our experiments we used data collected in Summer 2006 by crawling three blog hosting sites: Blogger, LiveJournal(LJ) and Xanga. The data consists of two main categories: user profiles containing various personal information provided by the user; and blog pages for recent entries in each crawled blog. We created an initial seed set of blogs and profiles by randomly identifying a subset of blogs hosted by each site. This initial seed set was expanded by downloading blogs (and corresponding profiles) referenced from the initial set. Our final data set therefore consists of (a subset of) blogs from each of the three crawled sites, corresponding profiles, and extracted links between blogs, and to webpages. Each web node corresponds to a single domain name (so links to `http://www.cnn.com/WEATHER/` and `http://www.cnn.com/US/` are counted as `www.cnn.com`). This improves the connectivity of the induced graph. We did not extract links from webpages back to blogs, since these were very rare. The results for the number of user profiles collected and the number of links extracted are shown in Table 4(c).

Demographics. We plotted the age, gender and location distribution of the users in our data set for each blog site. Ages claimed by blog users range from 1 to 120, with a small fraction of extreme values (on average, less than 0.6% with ages above 80). We did not filter implausible values, and this did not seem to impact results adversely. Not all users reveal their age, while some reveal only birthday, but not birth year. It can be observed from the distribution of ages in Figure 4(a) that Xanga has the youngest user population of the three networks, while Blogger has the most mature population among the three. Among the users that listed their gender, 63% of Xanga users are females while Blogger has only 47% of female users. LiveJournal does not make gender information available. Each of the three hosting sites we crawled offered a different format for specifying location. Blogger allows free text in location fields "City/Town" and "Region/State", and LiveJournal allows free text in "City"; as a result many users provide nonsensical values. Xanga organizes its users into "metros" according to geographical hierarchy (continent, country, city). Figure 4(b) shows the distribution of locations by continent. Additional details on the data collection process and data analysis is in [3].

Label Distribution. We analyzed the number of labeled neighbors in our multigraph for each of the nodes per data source, taking into account only incoming edges. In Blogger, the average number of labeled neighbors is 0.45 per node (most have 0). In Livejournal and Xanga the average is 7.5 and 3.1 respectively; the median is 2 in Livejournal and 1 in Xanga. Including friend links did not make a significant difference.

Implementation Issues. We implemented our algorithms in C++ and performed a detailed set of experiments to compare the performance of our methods on the blog data. In each task, the blog nodes are labeled with one of the three types of label: continuous (age), binary (gender), nominal (location). We also varied the multigraph by setting different weights for the link types, discussed in Section 3.3. For the Iterative local algorithm we set the number of iterations, s, to five, and the voting function to $\arg\max$ (plurality voting). For Global nearest neighbor, we used correlation coefficient as the similarity function, with weighting factor $\alpha = 0.5$ when including web nodes as features. In each experimental setting, we performed 10-fold cross validation, and report the average scores over the 10 runs: the set of labeled nodes is further divided into 10 subsets and evaluation is in turn performed on each subset using the remaining 9 for training. Across all experiments, the results were highly consistent: the standard deviation was less than 2% in each case. Although we run our experiments on the entire data set, we can evaluate only on the labeled subset (which is different for each label type).

4.2 Accuracy Evaluation

Age label. Figure 5 summarizes the various experiments performed while labeling the blog nodes with ages 1 to 120. We evaluate against the *stated* age in the blog profile. The features used by the two algorithm classes, Local Iterative and Global NN, are derived from the labels on the training set. We observe that with this information alone it is possible to attain an accurate labeling in blog data. Note that due to the graph structure, some nodes have empty feature vectors due to lack of links and, no matter how we propagate information, will never have any useful features with which to label. We therefore exclude such nodes from our evaluation.

Figure 5(a) shows the performance of the Local Iterative method for different accuracy levels from exact prediction to predicting within five years. The predictions for LiveJournal and Blogger show that with label data alone, it is possible to label with accuracy about 60% and 50% respectively within 3 years difference of the reported age. For the Xanga dataset, which is the most densely connected, we saw results that are appreciably much stronger: 88% prediction accuracy within 2 years off the reported age. Figure 5(b) shows a similar plot for Global NN algorithm. The prediction accuracy is not significantly different than the Local Iterative method. We observed that both methods tended to make accurate predictions for the same set of nodes.

Multigraph Edges. Other plots in Figure 5 compare the accuracy with the inclusion of additional edge types at unit weight. For LiveJournal, both the local and global methods benefit from using just the friend links (Figures 5(c) and 5(d)), suggesting that these edges conform more strongly to the hypotheses. The number of explicit friend links in Xanga is quite small which explains the decrease in the prediction accuracy when considering only friends (Blogger does not define friend links explicitly, so is not included here). We observe that including both the friend and blog edges often *reduces* the accuracy, showing the importance of different edge weights in the multigraph case.

Multigraph Nodes. We next analyzed the multigraph composed of both blog and web nodes, linked by blog and web links. There was no significant difference (less than 2%) in the prediction accuracy for age of bloggers when web nodes were considered as shown in Figures 5(e) and Figure 5(f). However, including web nodes in the analysis greatly improves the connectivity of the graph: Figure 5(g) shows that including web links significantly reduces the number of connected components. Consequently a larger fraction of nodes are given labels (Figure 5(h)), with little change in overall accuracy.

A side-effect of the iterative labeling is to assign age labels to websites ("pseudo-ages"). For sites with highly targeted audiences, the age reflects the age of the audience: in all data sets Facebook, a social networking site frequented by college students, is assigned an age of 21, while its high-school only counterpart is assigned an age around 17. The assigned age also reflects the demographic of the blog network: the USA Today website is consistently assigned lower ages than the Washington Post, but these age

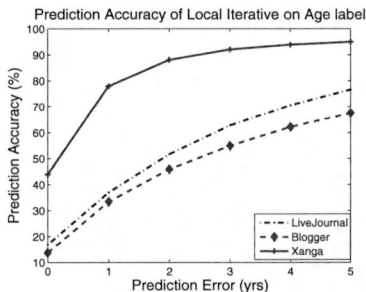

(a) Age with Local Iterative on Blog Links

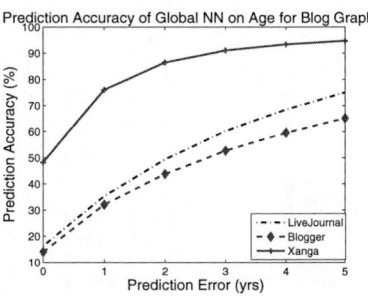

(b) Age with Global NN on Blog Links

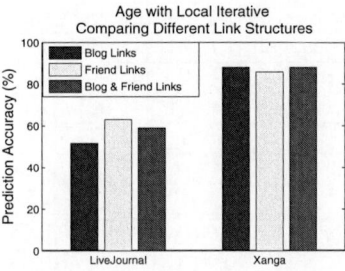

(c) Age with Local Iterative, Different Links

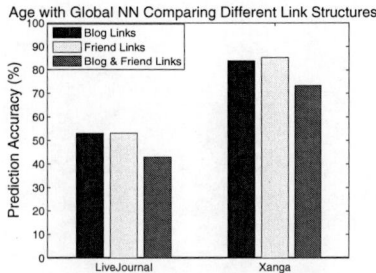

(d) Age with Global NN, Different Links

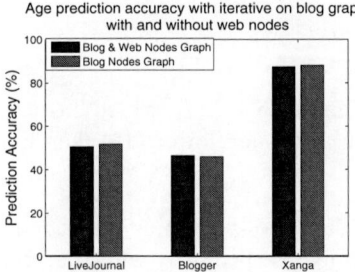

(e) Age with Local Iterative using Web Nodes

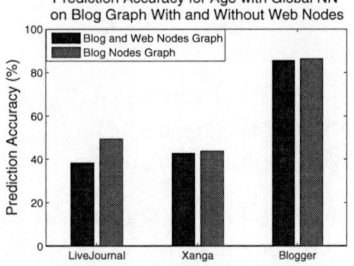

(f) Age with Global NN using Web Nodes

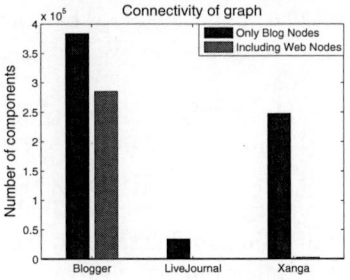

(g) Distribution of component size

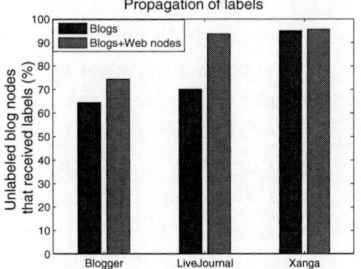

(h) Fraction of nodes assigned labels

Fig. 5. Experiments of Accuracy in Finding Age Labels, propagation results

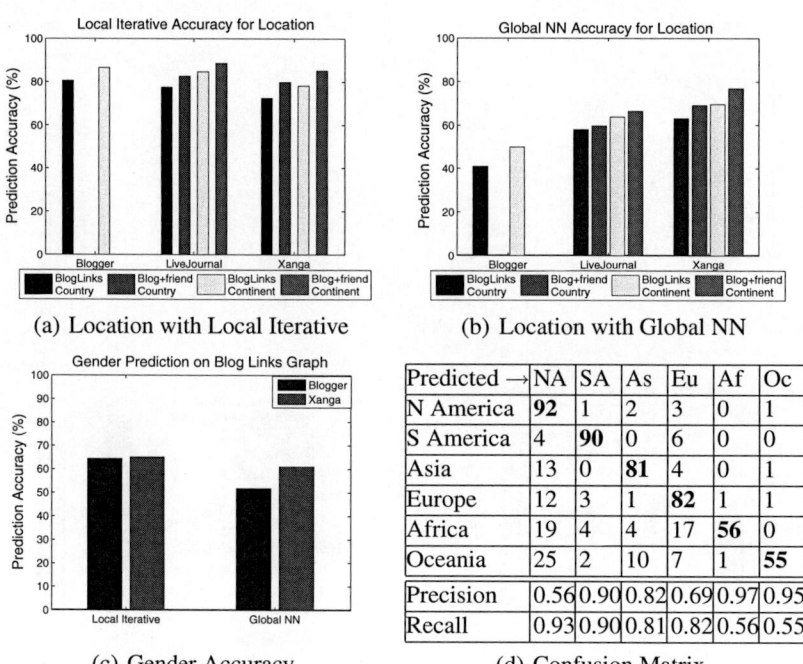

(a) Location with Local Iterative

(b) Location with Global NN

(c) Gender Accuracy

(d) Confusion Matrix

Predicted →	NA	SA	As	Eu	Af	Oc
N America	**92**	1	2	3	0	1
S America	4	**90**	0	6	0	0
Asia	13	0	**81**	4	0	1
Europe	12	3	1	**82**	1	1
Africa	19	4	4	17	**56**	0
Oceania	25	2	10	7	1	**55**
Precision	0.56	0.90	0.82	0.69	0.97	0.95
Recall	0.93	0.90	0.81	0.82	0.56	0.55

Fig. 6. Gender and Location Labels

pairs are higher in LiveJournal (28 vs. 25) than Xanga (23 vs. 17), which has younger users. The website for the band Slipknot is given age much lower (15) than that for Radiohead (28), reflecting the ages of their respective fans.

Location label. Figures 6(a) and 6(b) show the accuracy of predicting the country and continent of the blogger with local and global methods respectively while considering blog links and friend links. The local method predicts the country with about 80% accuracy for each dataset, significantly higher than the accuracy of predicting the correct age with no error. The reflects the fact that the hypothesis that connected blogs have similar locations (homophily) holds here. The global method performs less well, suggesting that the global hypothesis does not hold well for location.

For inferring the continent of the blogger, the local algorithm's accuracy is 88% for LiveJournal and about 85% for Blogger and Xanga (Figure 6(a)). This task should be easier than predicting country, and indeed all methods improve by up to 10 percentage points. Drilling down, the confusion matrix (in percentage) presented in Table 6(d) helps evaluate the performance of the classifier on the Blogger dataset with the local method. Analysis of precision and recall shows that the algorithm has a slight tendency to overrepresent common labels: North America (very common) has high recall but lower precision, while Africa (very rare) has high precision but lower recall.

Gender. As shown in Figure 6(c), our methods predict gender with up to 65% accuracy, with link information alone. This is better than random guessing, but not much, especially compared to methods that have looked at richer features (text content, word use, color schemes) [18]. Here, the local and global hypotheses do not hold well, and so more information is needed to improve the labeling.

4.3   Algorithm Performance Analysis

We compare the two methods by plotting the ROC curve for predicting the class of twenty-year old bloggers on the Xanga data set in Figure 7(a). The area under the curve (AUC) is 0.88 for Local Iterative and 0.87 for the Global NN method. Our methods give consistent accuracy as the fraction of labeled data for training is varied, even to below 1% of the total number of nodes (Figure 7(b)).

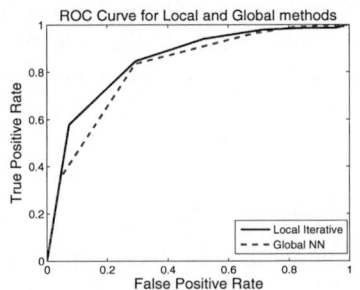

(a) ROC curve for Local and Global methods

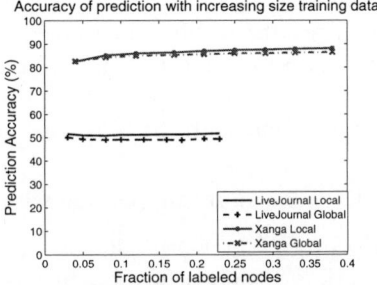

(b) Varying fraction of Labeled Data

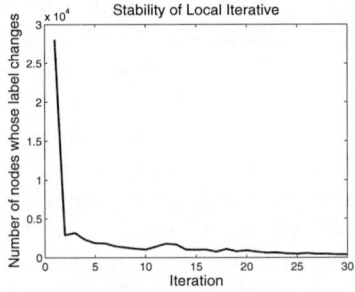

(c) Stability of Local Iterative

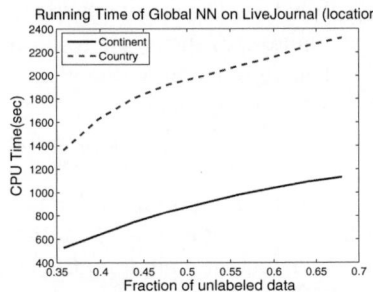

(d) Running Time of Global NN

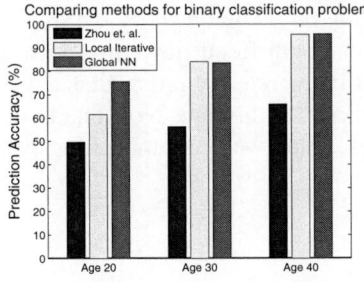

(e) Zhou *et al.* vs Local Iterative

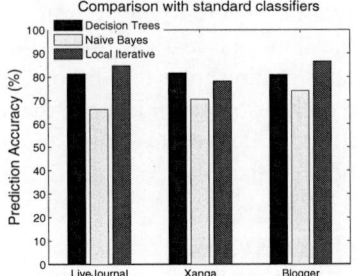

(f) Location label with standard classifiers

Fig. 7. Accuracy and Comparison with other methods

To study the impact of number of iterations for labeling, we studied the number of nodes whose age label *changes* during each iteration of the local method. The results for Blogger are plotted in Figure 7(c). There is a sharp decline in the first iteration, since many unlabeled nodes gain labels from their neighbors. The labeling quickly stabilizes, and although labels continue to shift, they do not impact the accuracy. We determined that just five iterations sufficed for a good quality labeling, and the accuracy was not very sensitive to this choice.

Recall that the running time of the global method depends on length of the feature vector, number of edges, labeled and unlabeled nodes from Theorem 2. Figure 7(d) shows the running time of the global method for varying feature vectors lengths and number of nodes and edges. The iterative algorithms (not shown) were significantly faster, typically taking around 10 seconds of CPU time instead of tens of minutes for the global case. The feature vector for countries is longer than that for continents, so continent labeling is faster than country labeling. Experiments on age (not shown) took longer, since the numbers of labeled and unlabeled nodes were approximately equal, the worst case for our algorithm. From the results of Figure 7(b), we could improve running time by using only a subset of the labeled nodes: that is, only compare to a (random) sample of the training data instead of every example.

4.4   Experimental Comparison with Previous Work

In prior work, Zhou *et al.* demonstrate a method for learning binary labels on graph data [24]. We implemented the method in Matlab using sparse matrix representation, with the same parameter values used in [24]. For comparison, we considered binary age labeling problems. The target label is whether the age is below a threshold of 20, 30, or 40 years. Figure 7(e) shows the result of ten-fold cross-validation on LiveJournal data: there is a benefit to using our local and global algorithms. The results were similar for the other datasets and labels.

We also compared with a large class of methods based on applying standard machine learning methods as a 'black box' over the feature vectors described above. This approach is at the core of prior work on relational learning and link classification [16,10]. We tested methods drawn from the Weka [20] library, and show only the *best* of these on location classification in Figure 7(f). Since the data is skewed, the Naïve Bayes classifier trivially assigns all test samples to one or two popular classes. The performance of Decision Trees approaches that of the local method. Analysis of these resulting trees shows that they seem to encode rules similar to the ones embodied by the local iterative algorithm: the learned trees rule out locations with frequency count zero, then use a series of thresholds to (approximately) identify the most common adjacent label. This strengthens our confidence in the local hypothesis for this data. Note that Decision Trees do not scale well to a large number of classes, and were significantly slower to learn and label than the local iterative approach.

5   Related Work

Our work is most closely related to problems of classification of relational data [7,19,16,6]. The area of Relational Learning is concerned with classification of objects that can be represented as relational databases. Our problem of graph labeling fits into this framework, since a (multi)graph can be encoded as tables of nodes and edges (one table for each node type, and one for each link type). Getoor *et al.* [7] proposed Probabilistic Relational Models (PRMs) which induce a Bayesian network over the given relational data to encode dependencies between each entity type and its attributes. This generative model is also limited by the constraint that the induced Bayesian network must be acyclic, which is typically not the case in blog/web networks: our graphs have cycles. To overcome this constraint, Taskar *et al.* [19,10] introduced Relational Markov Networks (RMN), a discriminative model based on Markov networks. Generalized Belief Propagation [21] is used for propagation of labels in the network. RMNs are described in the context of graphs with only one node type, instead of multigraphs, and rely on undirected edges—a limitation in settings where directionality provides information (as in web links). Neville *et al.* propose a decision tree variant Relational Probability Trees (RPT) [16], which incorporates aggregates over the neighborhood of a node (number of neighbors, average or mode of an attribute value, etc.).

Prior work often operates in the traditional supervised classification mode, with the assumption of a fully labeled graph to train on. This is not practical in our setting, especially where we observe a large graph of which only a subset of nodes are labeled. The induced fully labeled subgraph is typically so sparse that it is not representative, and may have few remaining links. Instead, one needs semi-supervised learning on graphs. Semi-supervised methods use the combination of labeled and unlabeled examples [25,26], viewed as a graph. Zhou *et al.* use graph regularization to impose a smoothness condition on the graph for labeling [24]. Their method is defined for a binary classification problem, and it not easily extensible to the general multi-class case. Earlier work [23] addresses the multi-class labeling problem, but does not consider the underlying link-structure in the data and assumes a distance-based affinity matrix. More recently, Zhang *et al.* [22] studied graph regularization for web-page categorization.

Our problem of graph labeling lies at the intersection of relational learning and semi-supervised learning. Here, prior approaches involve preprocessing the data such that it can be used as input to a known machine learning methods like random forests [2], or logistic regression [10]. This achieved by transforming the graph features into object attributes, and summarizing a multiset of neighboring nodes and their attributes by aggregate functions such as mean, mode, and count. Neville and Jensen [15] introduced the notion of an 'iterative classifier' over simple graphs (graphs with only one node type), which allows the result of one iteration of classification to influence the features used by the next round, for a fixed number of rounds. Similar to our local method, Macskassy *et al.* [12] proposed a simple Relational Neighbor (RN) classifier. The authors report comparable empirical performance of RN classifier and more sophisticated methods like PRM and RPT. Our graph labeling problem hearkens back to work of Chakrabarti *et al.* [5]. They observe that topic classification of webpages based on text can be improved by including information about the class of neighbors. Our aim is to go further and perform the classification based *only* on the neighborhood information from the link structure.

Analysis of blogs and other social media have been the focus of much research in the recent years. For classification in particular, prior work has studied blogs with respect to political orientation [1], mood [14], and so on. Non-link aware techniques like Natural Language Processing have been used for this [17]. There has also been some initial work on predicting the age and gender [18] and [4] of blog authors using text features. These papers showed the quality of certain textual features for producing similar labels to those we study here. Here, we show that links alone can be very powerful. More recently, the study by MacKinnon *et al.* [11] determined the probability distribution of friends in LiveJournal to infer location and age. The work of Hu *et al.* [8] uses a Bayesian framework to model web-click data for predicting age and gender; a side effect is to assign demographic attributes to the web pages themselves. Since click data is used, not link data, it applies to very different settings to those we consider.

6   Concluding Remarks

We have formalized the graph labeling problem for classification on blogs, and studied two classes of algorithms. These algorithms scale to large graphs, with hundreds of thousands of nodes and edges in a matter or minutes or seconds. On a case study with blog data, we see accuracy of up to 80-90% for correctly assigning labels, based only on link information. These results hold with a training set of 1% or even less compared to the size of the whole graph, and training data extracted automatically from profiles. It remains to validate these result on other domains, and to understand better how incorporating additional features can improve the results of these methods.

Acknowledgments. Supported by DyDAn, ONR grant number N00014-07-1-0150. and KDD supplement to NSF ITR 0220280.

References

1. Adamic, L.A., Glance, N.: The political blogosphere and the 2004 U.S. election: divided they blog. In: International Workshop on Link Discovery (LinkKDD), pp. 36–43 (2005)
2. Van Assche, A., Vens, C., Blockeel, H., Džeroski, S.: A random forest approach to relational learning. In: Workshop on Statistical Relational Learning (2004)
3. Bhagat, S., Cormode, G., Muthukrishnan, S., Rozenbaum, I., Xue, H.: No blog is an island - analyzing connections across information networks. In: International Conference on Weblogs and Social Media (2007)
4. Burger, J.D., Henderson, J.C.: Barely legal writers: An exploration of features for predicting blogger age. In: AAAI Spring Symposium on Computational Approaches to Analyzing Weblogs (2006)
5. Chakrabarti, S., Dom, B., Indyk, P.: Enhanced hypertext categorization using hyperlinks. In: ACM SIGMOD (1998)
6. Domingos, P., Richardson, M.: Markov logic: A unifying framework for statistical relational learning. In: Workshop on Statistical Relational Learning (2004)
7. Getoor, L., Friedman, N., Koller, D., Taskar, B.: Learning probabilistic models of link structure. Journal of Machine Learning Research 3, 679–707 (2002)
8. Hu, J., Zeng, H.-J., Li, H., Niu, C., Chen, Z.: Demographic prediction based on user's browsing behavior. In: International World Wide Web Conference (2007)
9. Indyk, P., Motwani, R.: Approximate nearest neighbors: Towards removing the curse of dimensionality. In: STOC (1998)

10. Lu, Q., Getoor, L.: Link-based classification. In: International Conference on Machine Learning (2003)
11. MacKinnon, I., Warren, R.H.: Age and geographic inferences of the LiveJournal social network. In: Statistical Network Analysis Workshop (2006)
12. Macskassy, S.A., Provost, F.: A simple relational classifier. In: Workshop on Multi-Relational Data Mining (2003)
13. McPherson, M., Smith-Lovin, L., Cook, J.M.: Birds of a feather: Homophily in social networks. Annual Review of Sociology 27, 415–444 (2001)
14. Mishne, G.: Experiments with mood classification in blog posts. In: Workshop on Stylistic Analysis of Text for Information Access (2005)
15. Neville, J., Jensen, D.: Iterative Classification in Relational Data. In: Workshop on Learning Statistical Models from Relational Data (2000)
16. Neville, J., Jensen, D., Friedland, L., Hay, M.: Learning relational probability trees. In: ACM Conference on Knowledge Discovery and Data Mining (SIGKDD) (2003)
17. Qu, H., Pietra, A.L., Poon, S.: Classifying blogs using NLP: Challenges and pitfalls. In: AAAI Spring Symposium on Computational Approaches to Analyzing Weblogs (2006)
18. Schler, J., Koppel, M., Argamon, S., Pennebaker, J.: Effects of age and gender on blogging. In: AAAI Spring Symposium on Computational Approaches to Analyzing Weblogs (2006)
19. Taskar, B., Abbeel, P., Koller, D.: Discriminative probabilistic models for relational data. In: Conference on Uncertainty in Artificial Intelligence (2002)
20. Witten, I.H., Frank, E.: Data Mining: Practical machine learning tools and techniques. Morgan Kaufmann, San Francisco (2005)
21. Yedidia, J., Freeman, W., Weiss, Y.: Generalized belief propagation. In: Advances in Neural Information Processing Systems (NIPS) (2000)
22. Zhang, T., Popescul, A., Dom, B.: Linear prediction models with graph regularization for web-page categorization. In: ACM Conference on Knowledge Discovery and Data Mining (SIGKDD) (2006)
23. Zhou, D., Bousquet, O., Lal, T.N., Weston, J., Schölkopf, B.: Learning with local and global consistency. In: Advances in Neural Information Processing Systems (2004)
24. Zhou, D., Huang, J., Schölkopf, B.: Learning from labeled and unlabeled data on a directed graph. In: International Conference on Machine Learning, pp. 1041–1048 (2005)
25. Zhu, X.: Semi-supervised learning literature survey. Technical report, Computer Sciences, University of Wisconsin-Madison (2006)
26. Zhu, X., Ghahramani, Z., Lafferty, J.: Semi-supervised learning using Gaussian fields and harmonic functions. In: International Conference on Machine Learning (2003)

Why We Twitter:
An Analysis of a Microblogging Community*

Akshay Java[1], Xiaodan Song[2], Tim Finin[1], and Belle Tseng[3]

[1] University of Maryland, Baltimore County
1000 Hilltop Circle
Baltimore MD 21250, USA
{aks1,finin}@cs.umbc.edu
[2] Google Inc.
1600 Amphitheatre Parkway
Mountain View, CA 94043, USA
xiaodansong@google.com
[3] Yahoo! Inc.
2821 Mission College Blvd
Santa Clara, CA 95054, USA
belle@yahoo-inc.com

Abstract. Microblogging is a new form of communication in which users describe their current status in short posts distributed by instant messages, mobile phones, email or the Web. We present our observations of the microblogging phenomena by studying the topological and geographical properties of the social network in Twitter, one of the most popular microblogging systems. We find that people use microblogging primarily to talk about their daily activities and to seek or share information. We present a taxonomy characterizing the the underlying intentions users have in making microblogging posts. By aggregating the apparent intentions of users in implicit communities extracted from the data, we show that users with similar intentions connect with each other.

1   Introduction

Microblogging is a variation on blogging in which users write short posts to a special blog that are subsequently distributed to their friends and other observers via text messaging, instant messaging systems, and email. Microblogging systems first appeared in mid 2006 with the launch of Twitter[1] and have multiplied to include many other systems, including Jaiku[2], Pownce[3], and others. These systems are generally seen as part of the "Web 2.0" wave of applications [1] and are still evolving.

---

* ©ACM, 2007. This is a minor revision of the work published in the Proceedings of the 9th WebKDD and 1st SNA-KDD 2007 workshop on Web mining and social network analysis. http://doi.acm.org/10.1145/1348549.1348556
[1] http://www.twitter.com
[2] http://www.jaiku.com
[3] http://www.pownce.com

H. Zhang et al. (Eds.): WebKDD/SNA-KDD 2007, LNCS 5439, pp. 118–138, 2009.
© Springer-Verlag Berlin Heidelberg 2009

Fig. 1. A Twitter member's recent microblogging posts can be viewed on the web along with a portion of the member's social network. This example shows the page for the first author with posts talking about his daily activities, thoughts and experiences.

Microblogging systems provide a light-weight, easy form of communication that enables users to broadcast and share information about their current activities, thoughts, opinions and status. One of the popular microblogging platforms is Twitter [2]. According to ComScore, within eight months of its launch, Twitter had about 94,000 users as of April, 2007 [3]. Figure 1 shows a snapshot of the first author's Twitter homepage. Updates or posts are made by succinctly describing one's current status through a short message (known in Twitter as a *tweet*) that is limited to 140 characters. Topics range from daily life to current events, news stories, and other interests. Instant messaging (IM) tools including Gtalk, Yahoo and MSN have features that allow users to share their current status with friends on their buddy lists. Microblogging tools facilitate easily sharing status messages either publicly or within a social network.

Microblogging differs from conventional blogging in both how and why people use it. Compared to regular blogging, microblogging fulfills a need for a faster and more immediate mode of communication. In constraining posts to be short enough to be carried by a single SMS (Short Message Service) message, microblogging systems also lower a user's investment of the time and thought required to generate the content. This also makes it feasible to generate the content on the limited keypads of mobile phones. The reduced posting burden encourages more frequent posting – a prolific blogger may update her blog every few days whereas a microblogger might post every few hours. The lowered barrier also supports new communication modes, including what one social media researcher [4] calls ambient intimacy.

> *Ambient intimacy is about being able to keep in touch with people with a level of regularity and intimacy that you wouldn't usually have access to, because time and space conspire to make it impossible.*

While the content of such posts ("I'm having oatmeal with raisins for breakfast") might seem trivial and unimportant, they are valued by friends and family members.

With the recent popularity of microblogging systems like Twitter, it is important to better understand *why* and *how* people use these tools. Understanding this will help us evolve the microblogging idea and improve both microblogging client and infrastructure software. We tackle this problem by studying the microblogging phenomena and analyzing different types of user intentions in such systems.

Much of research in user intention detection has focused on understanding the intent of a search queries. According to Broder [5], the three main categories of search queries are navigational, informational and transactional. Navigational queries are those that are performed with the intention of reaching a particular site. Informational queries are used to find resources on the Web such as articles, explanations etc. Transactional queries serve to locate shopping or download sites where further interactions would take place. According to this survey, most queries on the Web are informational or transactional in nature. Kang and Kim [6] present similar results in a study using a TREC data collection.

Understanding the intention for a search query is very different from user intention for content creation. In a survey of bloggers, Nardi et al. [7] describe different motivations for "why we blog". Their findings indicate that blogs are used as a tool to share daily experiences, opinions and commentary. Based on their interviews, they also describe how bloggers form communities online that may support different social groups in real world. Lento et al. [8] examined the importance of social relationship in determining if users would remain active users of the Wallop blogging system. A user's retention and interest in blogging was predicted by the comments received and continued relationship with other active members of the community. Users who are invited by people with whom they share pre-exiting social relationships tend to stay longer and active in the network. Moreover, certain communities were found to have a greater retention rate due to existence of such relationships. Mutual awareness in a social network has been found effective in discovering communities [9].

In computational linguistics, researchers have studied the problem of recognizing the communicative intentions that underlie utterances in dialog systems and spoken language interfaces. The foundations of this work go back to Austin [10], Stawson [11] and Grice [12]. Grosz [13] and Allen [14] carried out classic studies in analyzing the dialogues between people and between people and computers in cooperative task oriented environments. More recently, Matsubara [15] has applied intention recognition to improve the performance of automobile-based spoken dialog system. While their work focuses on the analysis of ongoing dialogs between two agents in a fairly well defined domain, studying user intention in Web-based systems requires looking at both the content and link structure.

In this paper, we describe how users have adopted the Twitter microblogging platform. Microblogging is relatively nascent, and to the best of our knowledge, no large scale studies have been done on this form of communication and information sharing. We study the topological and geographical structure of Twitter's social network and attempt to understand the user intentions and community structure in microblogging. Our analysis identifies four categories of microblogging intention: daily chatter, conversations, sharing information and reporting news. Furthermore, users play different roles of information source, friends or information seeker in different communities. We would like to discover what makes this environment so different and what needs are satisfied by such tools. In answering some of these questions, we also present a number of interesting statistical properties of user behavior and contrast them with blogging and other social network.

The paper is organized as follows: in Section 2, we describe the dataset and some of the properties of the underlying social network of Twitter users. Section 3 provides an analysis of Twitter's social network and its spread across geographies. Next, in Section 4 we describe aggregate user behavior and community level user intentions. Section 5 provides a taxonomy of user intentions. Finally, we summarize our findings and conclude with Section 6.

2   Dataset Description

Twitter is currently one of the most popular microblogging platforms. Users interact with this system by either using a Web interface, instant messaging agent or sending SMS messages. Members may choose to make their updates public or available only to friends. If user's profile is made public, her updates appear in a "public timeline" of recent updates and distributed to other users designated as friends or followers. The dataset used in this study was created by monitoring this public timeline for a period of two months, from April 01, 2007 to May 30, 2007. A set of recent updates were fetched once every 30 seconds. There are a total of 1,348,543 posts from 76,177 distinct users in this collection.

When we collected our data, Twitter's social network included two types of directed links between people: friend and follower. A Twitter user can "follow" another user, which results in their receiving notifications of public posts as they are made. Designating a twitter user as a friend also results in receiving post notifications, but indicates a closer relationship. The directed nature of both relations means that they can be one-way or reciprocated. The original motivation for having two relationships was privacy – a microblogger could specify the some posts were to be visible only to her friends and not to her (mere) followers. After the data was collected, Twitter changed its framework and eliminated the distinction, resulting in a single, directed relationship, follow, and a different mechanism for controlling who is notified about what posts.

By using the Twitter developer API[4], we fetched the social network of all users. We construct a directed graph $G(V, E)$, where V represents a set of users and E represents the set of "friend" relations. A directed edge e exists between two users u and v if user u declares v as a friend. There are a total of 87,897 distinct nodes with 829,053 friend relation between them. There are more nodes in this graph due to the fact that some users discovered though the link structure do not have any posts during the duration in which the data was collected. For each user, we also obtained their profile information and mapped their location to a geographic coordinate, details of which are provided in the following section.

3   Microblogging in Twitter

This section describes some of the characteristic properties of Twitter's social network, including its network topology, geographical distribution and common graph properties.

3.1   Growth of Twitter

Since Twitter provides a sequential user and post identifier, we can estimate the growth rate of Twitter. Figure 2 shows the growth rate for users and the Figure 3 shows the growth rate for posts in this collection. Since, we do not have access to historical data, we can only observe its growth for a two month time period. For each day we identify the maximum values for the user identifier and post identifier as provided by the Twitter API. By observing the change in these values, we can roughly estimate the growth of Twitter. It is interesting to note that even though Twitter launched in mid 2006, it really became popular soon after it won the South by SouthWest (SXSW) conference Web Awards[5] in March, 2007. Figure 2 shows the initial growth in users as a result of interest and publicity that Twitter generated at this conference. After this period, the rate at which new users are joining the network has declined. Despite the slow down, the number of new posts is constantly growing, approximately doubling every month indicating a steady base of users generating content.

Following Kolari et al. [16], we use the following definition of user activity and retention:

> **Definition.** A user is considered *active* during a week if he or she has posted at least one post during that week. **Definition.** An active user is considered *retained* for the given week, if he or she reposts at least once in the following X weeks.

Due to the short time period for which the data is available and the nature of microblogging, we chose as a value of X as a period of one week when computing user retention. Figure 4 shows the user activity and retention metrics for the

---

[4] http://twitter.com/help/api
[5] http://2007.sxsw.com/

Twitter Growth (Users)

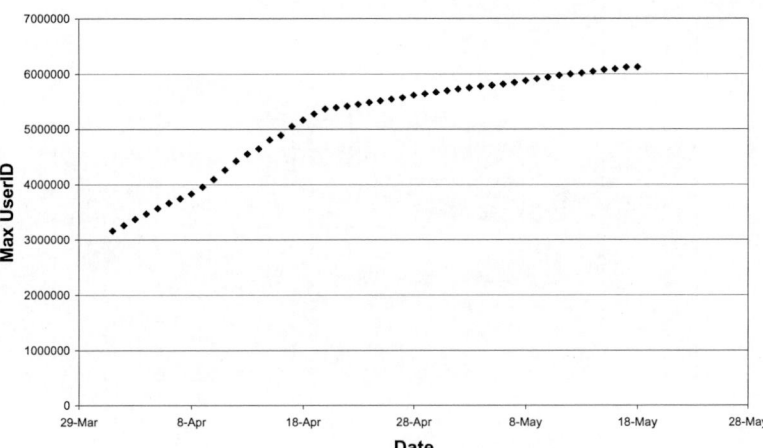

Fig. 2. During the time we collected data Twitter was growing rapidly. This figure shows the maximum userid observed for each day in the dataset. After an initial period of interest around March 2007, the rate at which new users are joining Twitter slowed.

Twitter Growth Rate (Posts)

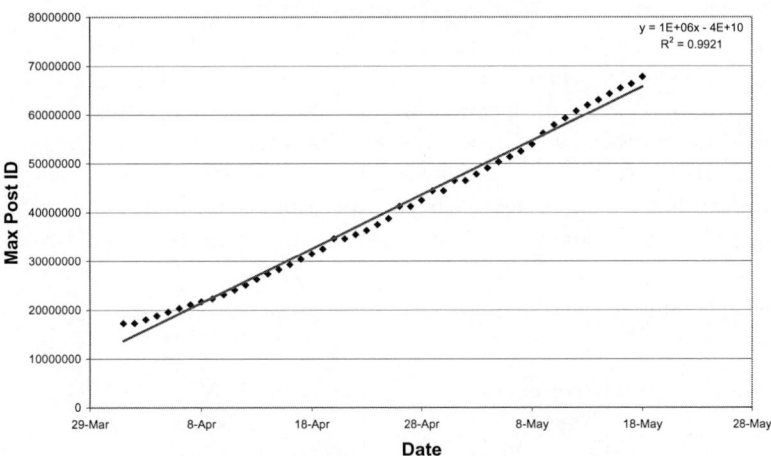

Fig. 3. During the data collection period the number of posts increased at a steady rate even as the rate at which new users joined slowed. This figure shows the maximum post ID observed for each day in the dataset.

duration of the data. About half of the users are active and of these, half of them repost in the following week. There is a lower activity recorded during the last week of the data due to the fact that updates from the public timeline are not available for two days during this period.

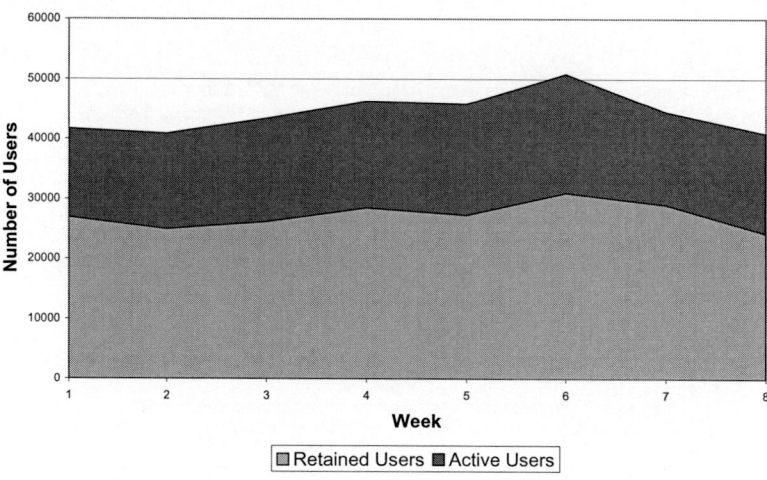

Fig. 4. The number of active and retained users remained fairly constant during the time the data was collected

3.2   Network Properties

The Web, Blogosphere, online social networks and human contact networks all belong to a class of "scale-free networks" [17] and exhibit a "small world phenomenon" [18]. It has been shown that many properties including the degree distributions on the Web follow a power law distribution [19,20]. Recent studies have confirmed that some of these properties also hold true for the Blogosphere [21].

Table 1 describes some of the properties for Twitter's social network. We also compare these properties with the corresponding values for the Weblogging Ecosystems Workshop (WWE) collection [22] as reported by Shi et al.

Table 1. This table shows the values of standard graph statistics for the Twitter social network

Property	Twitter	WWE
Total Nodes	87897	143,736
Total Links	829247	707,761
Average Degree	18.86	4.924
Indegree Slope	-2.4	-2.38
Outdegree Slope	-2.4	NA
Degree correlation	0.59	NA
Diameter	6	12
Largest WCC size	81769	107,916
Largest SCC size	42900	13,393
Clustering Coefficient	0.106	0.0632
Reciprocity	0.58	0.0329

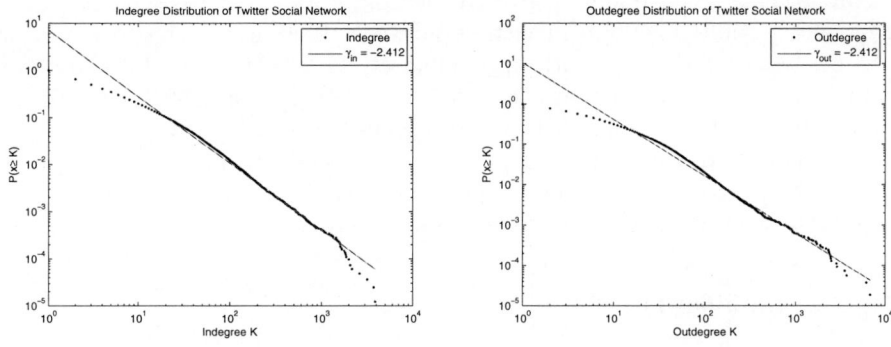

Fig. 5. The Twitter social network has a power law exponent of about -2.4, which is similar to value exhibited by the Web and Blogosphere

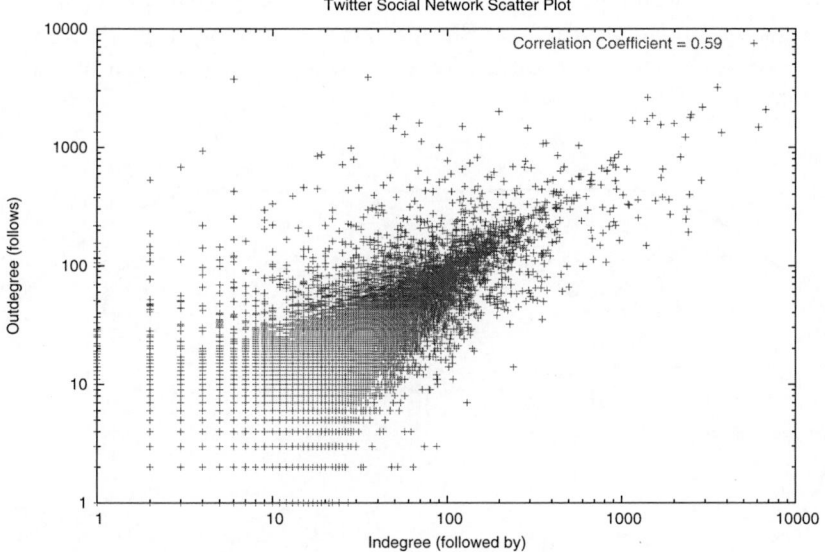

Fig. 6. This scatter plot shows the correlation between the indegree and outdegree for Twitter users. A high degree correlation signifies that users who are followed by many people also have large number of friends.

[21]. Their study shows a network with high degree correlation (also shown in Figure 6) and high reciprocity. This implies that there are a large number of mutual acquaintances in the graph. New Twitter users often initially join the network on invitation from friends. Further, new friends are added to the network by browsing through user profiles and adding other known acquaintances. High reciprocal links has also been observed in other online social networks like Livejournal [23]. Personal communication and contact network such as cell phone

call graphs [24] also have high degree correlation. Figure 5 shows the cumulative degree distributions [25,26] of Twitter's network. It is interesting to note that the slopes γ_{in} and γ_{out} are both approximately -2.4. This value for the power law exponent is similar to that found for the Web (typically -2.1 for indegree [27]) and Blogosphere (-2.38 for the WWE collection).

In terms of the degree distributions, Twitter's social network can thus be seen as being similar to the Web and Blogosphere, but in terms of reciprocity and degree correlation it is like a social network [23,24].

3.3   Geographical Distribution

Twitter provides limited profile information such as name, biographical sketch, timezone and location. For the 76 thousand users in our collection, slightly over half (about 39 thousand) had specified locations that could be parsed correctly and resolved to their respective latitude and longitudinal coordinates (using the Yahoo! Geocoding API[6]). Figure 7 and Table 2 show the geographical distribution of Twitter users and the number of users in each continent. Twitter is most popular in North America, Europe and Asia (mainly Japan). Tokyo, New York and San Francisco are the major cities where user adoption of Twitter is high [28].

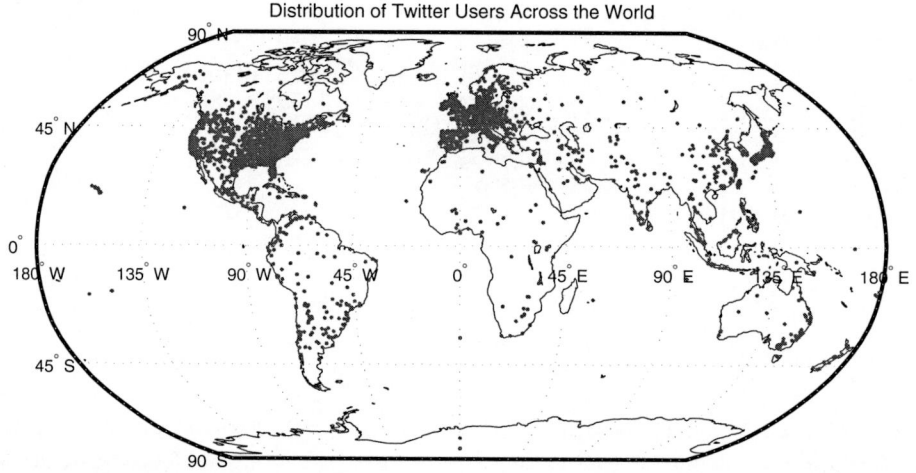

Fig. 7. Although Twitter was launched in United States, it is popular across the world. This map shows the distribution of Twitter users in our dataset.

Twitter's popularity is global and the social network of its users crosses continental boundaries. By mapping each user's latitude and longitude to a continent location we can extract the origin and destination location for every edge.

---

[6] http://developer.yahoo.com/maps/

Table 2. This table shows the geographical distribution of Twitter users, with North America, Europe and Asia exhibiting the highest adoption

Continent	Number of Users
North America	21064
Europe	7442
Asia	6753
Oceania	910
South America	816
Africa	120
Others	78
Unknown	38994

Table 3. This table shows the distribution of Twitter social network links across continents. Most of the social network lies within North America. (N.A = North America, S.A = South America)

from-to	Asia	Europe	Oceania	N.A	S.A	Africa
Asia	**13.45**	0.64	0.10	**5.97**	0.005	0.01
Europe	0.53	**9.48**	0.25	6.16	0.17	0.02
Oceania	0.13	0.40	0.60	1.92	0.02	0.01
N.A	**5.19**	**5.46**	1.23	**45.60**	0.60	0.10
S.A	0.06	0.26	0.02	0.75	0.62	0.00
Africa	0.01	0.03	0.00	0.11	0.00	0.03

Table 4. Comparing the social network properties within continents shows that Europe and Asia have a higher reciprocity indicating closer ties in these social networks. (N.A = North America)

Property	N.A	Europe	Asia
Total Nodes	16,998	5201	4886
Total Edges	205,197	42,664	60519
Average Degree	24.15	16.42	24.77
Degree Correlation	0.62	0.78	0.92
Clustering Coefficient	0.147	0.54	0.18
Percent Reciprocity	62.64	71.62	81.40

Table 3 shows the distribution of friendship relations across major continents represented in the dataset. Oceania is used to represent Australia, New Zealand and other island nations. A significant portion (about 45%) of the social network still lies within North America. Moreover, there are more intra-continent links than across continents. This is consistent with observations that the probability of friendship between two users is inversely proportionate to their geographic proximity [23].

Table 4 compares some of the network properties across these three continents with most users: North America, Europe and Asia. For each continent the social network is extracted by considering only the subgraph where both the source and destination of the friendship relation belong to the same continent. Asian and European communities have a higher degree correlation and reciprocity than their North American counterparts. Language plays an important role is such social networks. Many users from Japan and Spanish speaking world connect with others who speak the same language. In general, users in Europe and Asia tend to have higher reciprocity and clustering coefficient values in their corresponding subgraphs.

4 User Intention

Our analysis of user intention uses a two-level approach incorporating both HITS and community detection. First, we adapt the HITS algorithm [29] to find the hubs and authorities in the Twitter social network. An authority value for a person is the sum of the scaled hub values of her followers and her hub value is the sum of the scaled authority values of those she follows. Hubs and authorities have a mutually reinforcing property and are defined more formally as follows: $H(p)$ represents the hub value of the page p and $A(p)$ represents the authority value of a page p.

$$Authority(p) = \sum_{v \in S, v \to p} Hub(v)$$

And

$$Hub(p) = \sum_{u \in S, p \to u} Authority(u)$$

Table 5 shows a listing of Twitter users with the highest values as hubs and authorities. From this list, we can see that some users have high authority score, and also high hub score. For example, Scobleizer, JasonCalacanis, bloggersblog, and Webtickle who have many followers and friends in Twitter are located in

Table 5. This table lists the Twitter users with the top hub and authority values computed from our dataset. Some of the top authorities are also popular bloggers. Top hubs include users like startupmeme and aidg which are microblogging versions of a blogs and other Web sites.

User	Authority	User	Hub
Scobleizer	0.002354	Webtickle	0.003655
Twitterrific	0.001765	Scobleizer	0.002338
ev	0.001652	dan7	0.002079
JasonCalacanis	0.001557	startupmeme	0.001906
springnet	0.001525	aidg	0.001734
bloggersblog	0.001506	lisaw	0.001701
chrispirillo	0.001503	bhartzer	0.001599
darthvader	0.001367	bloggersblog	0.001559
ambermacarthur	0.001348	JasonCalacanis	0.001534

this category. Some users with very high authority scores have relatively low hub score, such as Twitterrific, ev, and springnet. They have many followers while less friends in Twitter, and thus are located in this category. Some other users with very high hub scores have relatively low authority scores, such as dan7, startupmeme, and aidg. They follow many other users while have less friends instead. Based on this rough categorization, we can see that user intention can be roughly categorized into these three types: information sharing, information seeking, and friendship-wise relationship.

After the hub/authority detection, we identify communities within friendship-wise relationships by only considering the bidirectional links where two users regard each other as friends. A community in a network can be defined as a group of nodes more densely connected to each other than to nodes outside the group. Often communities are topical or based on shared interests. To construct web communities Flake et al. [30] proposed a method using HITS and maximize flow/minimize cut to detect communities. In social network area, Newman and Girvan [31,32] proposed a metric called modularity to measure the strength of the community structure. The intuition is that a good division of a network into communities is not merely to make the number of edges running between communities small; rather, the number of edges between groups is smaller than expected. Only if the number of between group edges is significantly lower than what would be expected purely by chance can we justifiably claim to have found significant community structure. Based on the modularity measure of the network, optimization algorithms are proposed to find good divisions of a network into communities by optimizing the modularity over possible divisions. Also, this optimization process can be related to the eigenvectors of matrices. However, in the above algorithms, each node has to belong to one community, while in real networks, communities often overlap. One person can serve a totally different functionality in different communities. In an extreme case, one user can serve as the information source in one community and the information seeker in another community.

In this section we describe some specific examples of how communities form in Twitter. Communities are the building blocks of any social network tools. Often the communities that develop are topical or based on shared interests. A community in a network is a group of nodes more densely connected to each other than to nodes outside the group. In naturally occurring networks, of course, communities often overlap.

People in friendship communities often know each other. Prompted by this intuition, we applied the Clique Percolation Method (CPM) [33,34] to find overlapping communities in networks. The CPM is based on the observation that a typical member in a community is linked to many other members, but not necessarily to all other nodes in the same community. In CPM, the k-clique-communities are identified by looking for the unions of all k-cliques that can be reached from each other through a series of adjacent k-cliques, where two k-cliques are said to be adjacent if they share k-1 nodes. This algorithm is suitable for detecting the dense communities in the network.

Here we give some specific examples of implicit Twitter communities and characterize the apparent intentions that their users have in posting. These "community intentions" can provide insight into why the communities emerge and the motivations users have in joining them Figure 8 illustrates a representative community with 58 users closely communicating with each other through Twitter service. The key terms they talk about include work, Xbox, game, and play. It looks like some users with gaming interests getting together to discuss the information about certain new products on this topic or sharing gaming experience. When we go to specific users website, we also find the following type of conversation.

> "BDazzler@Steve519 I don't know about the Jap PS3's. I think they have region encoding, so you'd only be able to play Jap games. Euro has no ps2 chip" or "BobbyBlackwolf Playing with the PS3 firmware update, can't get WMP11 to share MP4's and the PS3 won't play WMV's or AVI's...Fail."

We also noticed that users in this community share with each other their personal feeling and daily life experiences in addition to comments on "gaming". Based on our study of the communities in Twitter dataset, we observed that this is a representative community in Twitter network: people in one community have certain common interests and they also share with each other about their personal feeling and daily experience.

Using the Clique Percolation Method we are able to find how communities are connected to each other by overlapping components. Figure 9 illustrates two communities with podcasting interests where the users *GSPN* and *pcamarata*

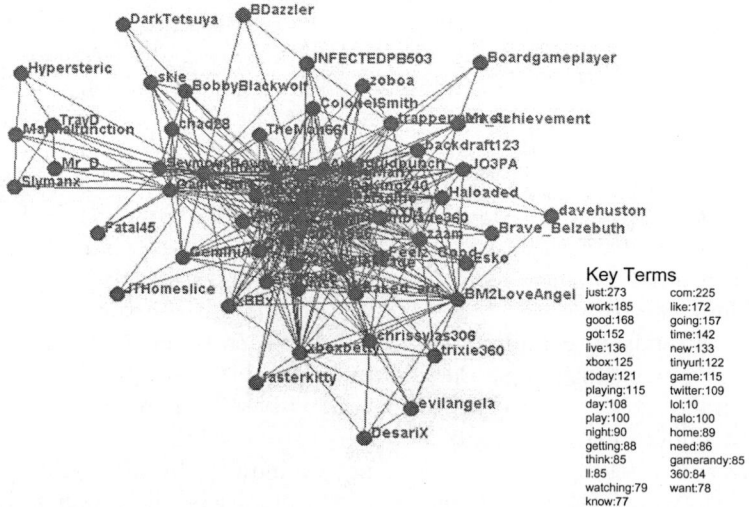

Fig. 8. One of the user communities we discovered in the Twitter dataset is characterized by an interest in computer games, which is the major topic of the community members' posts. As is typical of other topic-based communities, the members also use Twitter to share their daily activities and experiences.

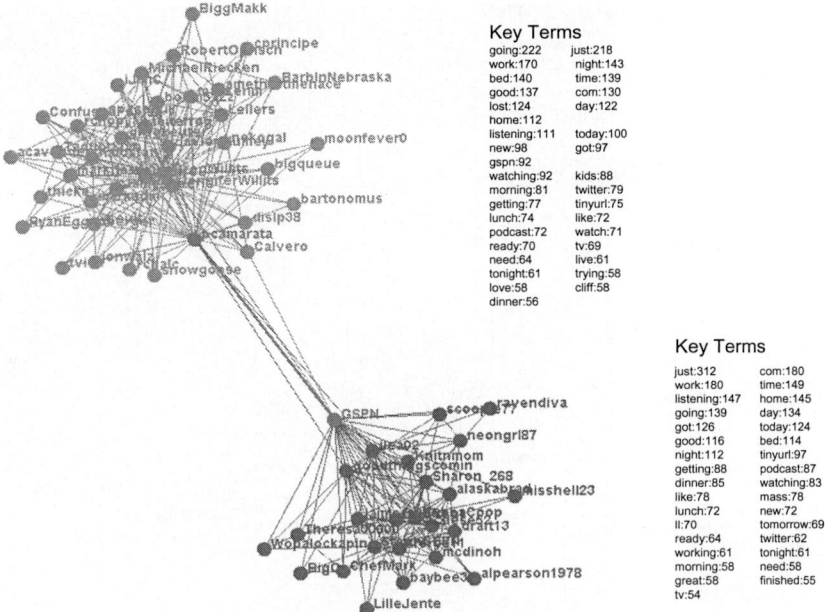

Fig. 9. Our analysis revealed two Twitter communities in which podcasting was a dominant topic that are connected by two individuals. The communities differ in topic diversity, with the red community having a narrower focus.

connect these two communities. In GSPN's biographic sketch, he states that he is the producer of the *Generally Speaking Porkiest Network*[7]; while in pcamarata's bio, he mentioned he is a family man, a neurosurgeon, and a podcaster. By looking at the top key terms of these two communities, we can see that the focus of the green community is a little more diversified: people occasionally talk about podcasting, while the topic of the red community is a little more focused. In a sense, the red community is like a professional community of podcasting while the green one is a informal community about podcasting.

Figure 11 shows two example communities whose members tend to talk about their daily activities and thoughts. These discussions, while may seem mundane to most of us, will be of interest to close friends and family members. It is very similar to keeping in touch with friends and maintaining regular contact with them on chat or instant messengers. As pointed out by Reichelt [4], this use of microblogging can create a sense of awareness and intimacy that transcends the constraints of space and time.

Figure 10 illustrates five communities connected by *Scobleizer*, who is well known as a blogger specializing in technology. People follow his posts to get technology news. People in different communities share different interests with

---

[7] http://ravenscraft.org/gspn/home

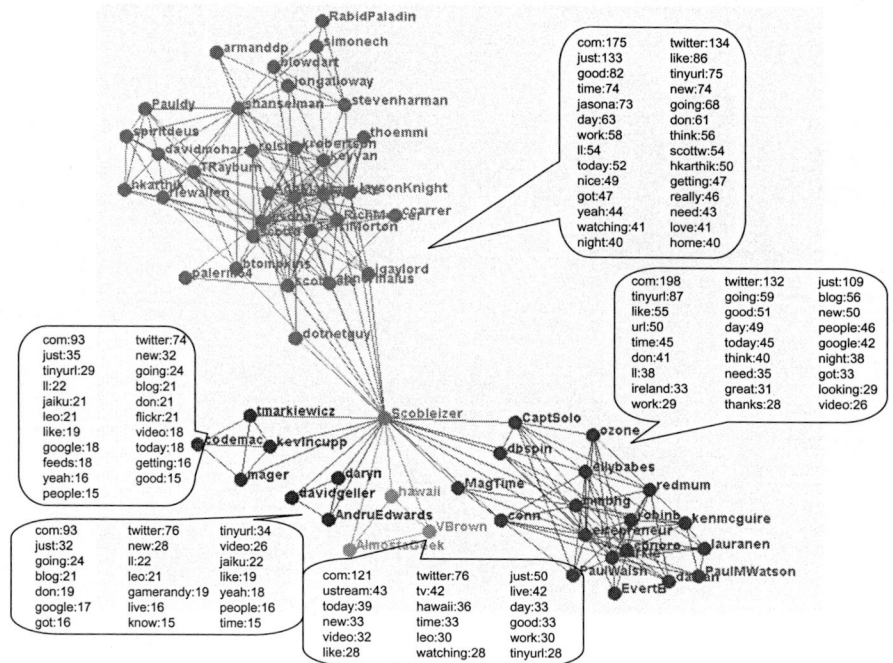

Fig. 10. We identified five communities sharing technology as a dominant topic that are connected by a single user – *Scobleizer*, well known as a blogger specializing in technology

Scobleizer. Specifically, the Twitter users *AndruEdwards*, *Scobleizer*, *daryn*, and *davidgeller* get together to share video related news. CaptSolo et al. have some interests on the topic of the Semantic Web. *AdoMatic* and others are engineers and have interests focused on computer programming and related topics.

Figure 9 shows how two seemingly unrelated communities can be connected to each other through a few weak ties [35]. While Twitter itself does not support any explicit communities, structures naturally emerge in the Twitter social network. Providing an easy way for users to find others in their implicit communities might be a useful service for systems like Twitter.

Studying intentions at a community level, we observe users participate in communities that share similar interests. Individuals may have different intentions in becoming a part these implicit communities. While some act as information providers, others are merely looking for new and interesting information. Next, we analyze aggregate trends across users spread over many communities, we can identify certain distinct themes. Often there are recurring patterns in word usages. Such patterns may be observed over a day or a week. For example Figure 12 shows the trends for the terms "friends" and "school" in the entire corpus. While school is of interest during weekdays, the frequency of the friends term increases during the week and dominates on the weekends.

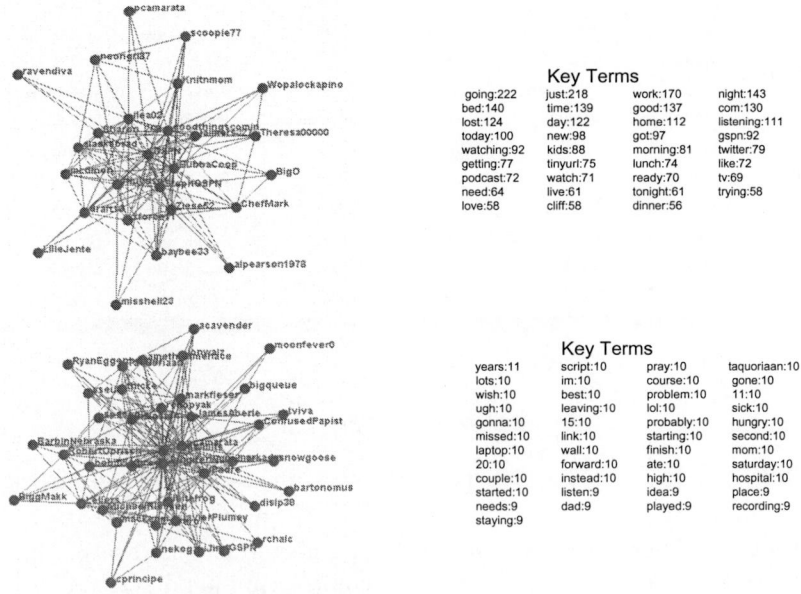

Fig. 11. Analyzing the key terms in these two communities show that their members post updates that discuss the events of their daily activities

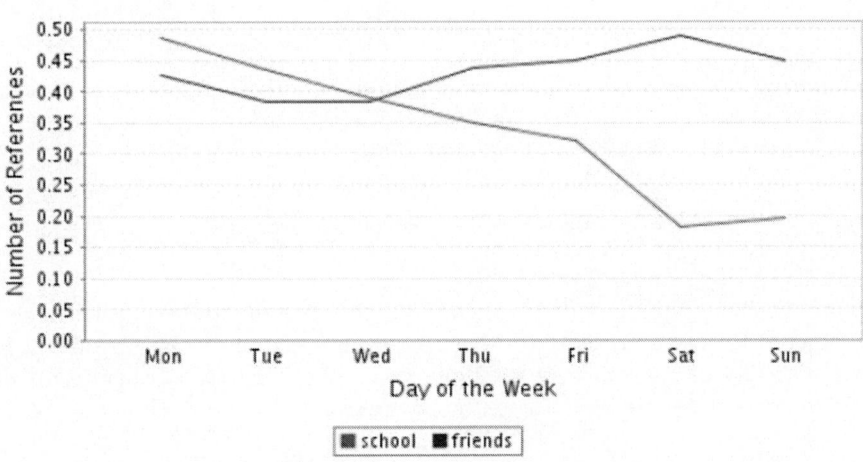

Fig. 12. This graph shows the daily trends for terms "school" and "friends". The term "school" is more frequent during the early week while "friends" take over during the weekend.

Monday	Heroes	Digg	Lost	Friday	Saturday	Easter
Week	Como	Net	Vonnegut	Weekend	**Watching**	Sunday
DRM	**Email**	Mac	**Meeting**	Office	Night	Happy
Emi	**24**	Wednesday	Thursday	TGIF	**Party**	**Watching**
Mondays	**Working**	Meeting	**Idol**	Viernes	**Playing**	Mother
April	**Class**	74 5b 9d..	Kurt	Tonight	Movie	**Church**
School	Tuesdays	Lunch	Tw	Good	By	Brunch
Sopranos	May	Sportsalert	**Working**	Here	**Shopping**	Family
News	**Meeting**	**Idol**	Lunch	Beer	Been	**Dinner**
Weekend	Funciona	Today	Office	**Go**	Jaiku	Mothers
Start	Joost	like	**American**	13th	Eurovision	**Playing**
Mon	**Tue**	**Wed**	**Thu**	**Fri**	**Sat**	**Sun**

TV Shows Current Event Activity

Fig. 13. Distinctive terms for each day of the week ranked using Log-likelihood ratio

The log-likelihood ratio is used to determine terms that are of significant importance for a given day of the week. Using a technique described by Rayson and Garside [36], we create a contingency table of term frequencies for each of the day of the week and the remaining days in the week.

	Day	Other Days	Total
Frequency of word	a	b	a+b
Frequency of other words	c-a	d-b	c+d-a-b
Total	c	d	c+d

Comparing the terms that occur on a given day with the histogram of terms for the rest of the week, we find the most descriptive terms. The log-likelihood score is calculated as follows:

$$LL = 2 * (a * log(\frac{a}{E1}) + b * log(\frac{b}{E2})) \tag{1}$$

where $E1 = c * \frac{a+b}{c+d}$ and $E2 = d * \frac{a+b}{c+d}$

Figure 13 shows the most descriptive terms for each day of the week. Some of the extracted terms correspond to recurring events and activities significant for a particular day of the week for example "school" or "party". Other terms are related to current events like "Easter" and "EMI".

5   Discussion

This section presents a brief taxonomy of user intentions in the Twitter microblogging community. The apparent intention of a Twitter post was determined manually by the first author. Each post was read and categorized. Posts that were highly ambiguous or for which the author could not make a judgement were placed in the category UNKNOWN. Based on this analysis we have found the following as the main user intentions in Twitter posts.

- *Daily Chatter.* Most posts on Twitter talk about daily routine or what people are currently doing. This is the largest and most common user of Twitter.
- *Conversations.* In Twitter, since there is no direct way for people to comment or reply to their friend's posts, early adopters started using the @ symbol followed by a username for replies. About one eighth of all posts in the collection contain a conversation and this form of communication was used by almost 21% of users in the collection.
- *Sharing information/URLs.* About 13% of all the posts in the collection contain some URL in them. Due to the small character limit of Twitter updates, a URL shortening service like TinyURL[8] is frequently used to make this feature feasible.
- *Reporting news.* Many users report latest news or comment about current events on Twitter. Some automated users or agents post updates like weather reports and new stories from RSS feeds. This is an interesting application of Twitter that has evolved due to easy access to the developer API.

Using the link structure, following are the main categories of users on Twitter:

- *Information Source.* An Twitter user who is an information source is also a hub and has a large number of followers. This user may post updates on regular intervals or infrequently. Despite infrequent updates, certain users have a large number of followers due to the valuable nature of their updates. Some of the information sources were also found to be automated tools posting news and other useful information on Twitter.
- *Friends.* Most relationships fall into this broad category. There are many subcategories of friendships on Twitter. For example a user may have friends, family and co-workers on their friend or follower lists. Sometimes unfamiliar users may also add someone as a friend.
- *Information Seeker.* An information seeker is a person who might post rarely, but follows other users regularly.

Our study has revealed different motivations and utilities of microblogging platforms. A single user may have multiple intentions or may even serve different roles in different communities. For example, there may be posts meant to update your personal network on a holiday plan or a post to share an interesting link with co-workers. Multiple user intentions have led to some users feeling overwhelmed by microblogging services [37]. Based on our analysis of user intentions, we believe that the ability to categorize friends into groups (e.g. family, co-workers) would greatly benefit the adoption of microblogging platforms. In addition features that could help facilitate conversations and sharing news would be beneficial.

---

[8] http://www.tinyurl.com

6   Conclusion

In this study we have analyzed a large social network in a new form of social media known as microblogging. Such networks were found to have a high degree correlation and reciprocity, indicating close mutual acquaintances among users. While determining an individual user's intention in using such applications is challenging, by analyzing the aggregate behavior across communities of users, we can describe the community intention. Understanding these intentions and learning *how* and *why* people use such tools can be helpful in improving them and adding new features that would retain more users.

We collected two months of data from the Twitter microblogging system, including information on users, their social networks and posts. We identified different types of user intentions and studied the community structures. Our ongoing work includes the development of automated approaches of detecting user intentions with related community structures and the design of efficient techniques to extract community structures from very large social networks [38].

Acknowledgements

We would like to thank Twitter Inc. for providing an API to their service and Pranam Kolari, Xiaolin Shi and Amit Karandikar for their suggestions.

References

1. Griswold, W.G.: Five enablers for mobile 2.0. Computer 40(10), 96–98 (2007)
2. Pontin, J.: From many tweets, one loud voice on the internet. The New York Times (April 22, 2007)
3. Comscore: Sites for social butterflies (May 2007), http://www.usatoday.com/tech/webguide/2007-05-28-social-sites_N.htm
4. Reichelt, L.: (March 2007), http://www.disambiguity.com/ambient-intimacy/
5. Broder, A.: A taxonomy of web search. SIGIR Forum 36(2), 3–10 (2002)
6. Kang, I.H., Kim, G.: Query type classification for web document retrieval. In: SIGIR 2003: Proceedings of the 26th annual international ACM SIGIR conference on Research and development in informaion retrieval, pp. 64–71. ACM Press, New York (2003)
7. Nardi, B.A., Schiano, D.J., Gumbrecht, M., Swartz, L.: Why we blog. Commun. ACM 47(12), 41–46 (2004)
8. Lento, T., Welser, H.T., Gu, L., Smith, M.: The ties that blog: Examining the relationship between social ties and continued participation in the wallop weblogging system (2006)
9. Lin, Y.R., Sundaram, H., Chi, Y., Tatemura, J., Tseng, B.: Discovery of Blog Communities based on Mutual Awareness. In: Proceedings of the 3rd Annual Workshop on Weblogging Ecosystem: Aggregation, Analysis and Dynamics, 15th World Wide Web Conference (May 2006)
10. Austin, J.: How to Do Things with Words. Oxford University Press, Oxford (1976)
11. Strawson, P.: Intention and Convention in Speech Acts. The Philosophical Review 73(4), 439–460 (1964)

12. Grice, H.: Utterer's meaning and intentions. Philosophical Review 78(2), 147–177 (1969)
13. Grosz, B.J.: Focusing and Description in Natural Language Dialogues. Cambridge University Press, New York (1981)
14. Allen, J.: Recognizing intentions from natural language utterances. In: Computational Models of Discourse, 107–166 (1983)
15. Matsubara, S., Kimura, S:, Kawaguchi, N., Yamaguchi, Y., Inagaki, Y.: Example-based Speech Intention Understanding and Its Application to In-Car Spoken Dialogue System. In: Proceedings of the 19th international conference on Computational linguistics, vol. 1, pp. 1–7 (2002)
16. Kolari, P., Finin, T., Yesha, Y., Yesha, Y., Lyons, K., Perelgut, S., Hawkins, J.: On the Structure, Properties and Utility of Internal Corporate Blogs. In: Proceedings of the International Conference on Weblogs and Social Media (ICWSM 2007) (March 2007)
17. Barabasi, A.L., Albert, R.: Emergence of scaling in random networks. Science 286, 509 (1999)
18. Watts, D.J., Strogatz, S.H.: Collective dynamics of 'small-world' networks. Nature 393(6684), 440–442 (1998)
19. Kumar, R., Raghavan, P., Rajagopalan, S., Tomkins, A.: Trawling the Web for emerging cyber-communities. Computer Networks 31(11-16), 1481–1493 (1999)
20. Broder, A., Kumar, R., Maghoul, F., Raghavan, P., Rajagopalan, S., Stata, R., Tomkins, A., Wiener, J.: Graph structure in the web. In: Proceedings of the 9th international World Wide Web conference on Computer networks: the international journal of computer and telecommunications networking, pp. 309–320. North-Holland Publishing Co., Amsterdam (2000)
21. Shi, X., Tseng, B., Adamic, L.A.: Looking at the blogosphere topology through different lenses. In: Proceedings of the International Conference on Weblogs and Social Media (ICWSM 2007) (2007)
22. Blogpulse: The 3rd annual workshop on weblogging ecosystem: Aggregation, analysis and dynamics. In: 15th world wide web conference (May 2006)
23. Liben-Nowell, D., Novak, J., Kumar, R., Raghavan, P., Tomkins, A.: Geographic routing in social networks. Proceedings of the National Academy of Sciences 102(33), 11623–11628 (2005)
24. Nanavati, A.A., Gurumurthy, S., Das, G., Chakraborty, D., Dasgupta, K., Mukherjea, S., Joshi, A.: On the structural properties of massive telecom call graphs: findings and implications. In: CIKM 2006: Proceedings of the 15th ACM international conference on Information and knowledge management, pp. 435–444. ACM Press, New York (2006)
25. Newman, M.E.J.: Power laws, pareto distributions and zipf's law. Contemporary Physics 46, 323 (2005)
26. Clauset, A., Shalizi, C.R., Newman, M.E.J.: Power-law distributions in empirical data (June 2007)
27. Donato, D., Laura, L., Leonardi, S., Millozzi, S.: Large scale properties of the webgraph. European Physical Journal B 38, 239–243 (2004)
28. Java, A.: Global distribution of twitter users, http://ebiquity.umbc.edu/blogger/2007/04/15/global-distribution-of-twitter-users/
29. Kleinberg, J.M.: Authoritative sources in a hyperlinked environment. Journal of the ACM 46(5), 604–632 (1999)
30. Flake, G.W., Lawrence, S., Giles, C.L., Coetzee, F.: Self-organization of the web and identification of communities. IEEE Computer 35(3), 66–71 (2002)

31. Girvan, M., Newman, M.E.J.: Community structure in social and biological networks (December 2001)
32. Clauset, A., Newman, M.E.J., Moore, C.: Finding community structure in very large networks. Physical Review E 70, 066111 (2004)
33. Palla, G., Derenyi, I., Farkas, I., Vicsek, T.: Uncovering the overlapping community structure of complex networks in nature and society. Nature 435, 814 (2005)
34. Derenyi, I., Palla, G., Vicsek, T.: Clique percolation in random networks. Physical Review Letters 94, 160202 (2005)
35. Granovetter, M.S.: The strength of weak ties. The American Journal of Sociology 78(6), 1360–1380 (1973)
36. Rayson, P., Garside, R.: Comparing corpora using frequency profiling (2000)
37. Lavallee, A.: Friends swap twitters, and frustration - new real-time messaging services overwhelm some users with mundane updates from friends (March 16, 2007)
38. Java, A., Joshi, A., Finin, T.: Approximating the community structure of the long tail. In: The Second International Conference on Weblogs and Social Media (submitted) (November 2007)

A Recommender System Based on Local Random Walks and Spectral Methods

Zeinab Abbassi[1] and Vahab S. Mirrokni[2]

[1] Department of Computer Science, University of British Columbia,
201 2366 Main Mall, Vancouver, BC, Canada V6T1Z4
zeinab@cs.ubc.ca
[2] Microsoft Research, Redmond, WA, USA 98052
mirrokni@theory.csail.mit.edu

Abstract. In this paper, we design recommender systems for blogs based on the link structure among them. We propose algorithms based on refined random walks and spectral methods. First, we observe the use of the personalized page rank vector to capture the relevance among nodes in a social network. We apply the local partitioning algorithms based on refined random walks to approximate the personalized page rank vector, and extend these ideas from undirected graphs to directed graphs. Moreover, inspired by ideas from spectral clustering, we design a similarity metric among nodes of a social network using the eigenvalues and eigenvectors of a normalized adjacency matrix of the social network graph. In order to evaluate these algorithms, we crawled a set of blogs and construct a blog graph. We expect that these algorithms based on the link structure perform very well for blogs, since the average degree of nodes in the blog graph is large. Finally, we compare the performance of our algorithms on this data set. In particular, the acceptable performance of our algorithms on this data set justifies the use of a link-based recommender system for social networks with large average degree.[1]

1 Introduction

Recommender systems use the opinions of a community of users to help individuals in that community more effectively identify content of interest from a gigantic set of choices [3]. In social networks, recommender systems can be defined using the link structure among nodes of the network. This method is particularly useful for social networks with large average degree. One of the main social networks with high average degree is the network of blogs.

The rapid development of blogging is one of the most interesting phenomena in the evolution of the web over the last few years. As of December 2006, the web contains over 37 million blogs, with roughly 40,000 new blogs and 750,000

H. Zhang et al. (Eds.): WebKDD/SNA-KDD 2007, LNCS 5439, pp. 139–153, 2009.
© Springer-Verlag Berlin Heidelberg 2009

new posts created daily[1]. These blogs span a wide range, from personal journals read mainly by friends, to very popular blogs visited by millions of readers. All together, these blogs form a distinct, human-generated subset of the web (blogspace) which is increasingly valuable as a source of information. A comprehensive understanding of blogs and blogspace is likely to prove critical, and may also offer useful insights into such development as the behavior of social networks or information diffusion. On the other hand, the increasing amount of choices available on the Web has made it increasingly difficult to find what one is looking for. Three methods are commonly applied to help web users in locating their desired items: search engines, taxonomies and, more recently, recommender systems [2].

The problem we are addressing in this paper is to design an algorithm to recommend blog users what other blogs they may enjoy reading. To do so, we assume that each user has a set of *favorite blogs* and we want to identify a set of *similar* blogs to the favorite blogs. The blog recommender systems discussed in this paper can be applied to any social network in which the relevance and relations among nodes of the network are captured by links among nodes of a graph. In particular, these link-based methods are useful for social networks with high average degree.

1.1 Our Contribution

In this paper, we design, implement, and compare algorithms for recommender systems for social networks based on their link structure, and evaluate our algorithms on a data set of blogs. We first crawl a set of blogs, and constructed the link structure among them as a blog graph. We use ideas from spectral clustering algorithms [8] and a variant of local clustering algorithm based on refined random walks [14] and personalized PageRank algorithms [4] for our recommender system. In particular,

- We extend the truncated random walk algorithm to approximate the personalized PageRank vector [4, 14] from undirected graphs to directed graphs. To avoid accumulating probabilities on sinks in directed graphs, we design a non-uniform random walk as follows: we find strongly connected components of the directed blog graph, and change the restarting probability of the random walk based on the position of each node in the structure of the strongly connected component decomposition. This will result in a non-uniform random walk for directed graphs that gives better results compared to performing a uniform random walk on the directed graph. The details of this algorithm can be found in Section 3.1.
- We define a spectral algorithm in which we calculate a *similarity metric* among nodes of the blog graph based on the eigenvalues and eigenvectors of a normalized adjacency matrix of the blog graph. For a set of favorite weblogs, this algorithm outputs the closest nodes to the favorite weblogs according to the similarity metric. The details of this similarity metric can be found in Section 4.

– We analyze the structure of the blog graph for its average degree, and for its connected and strongly connected components. We then implement and evaluate the following four algorithms: a spectral algorithm applied to two candidate sets from the directed and undirected graph, and a truncated random walk algorithm to approximate the personalized PageRank vector on directed and undirected graphs. We compare these algorithms in terms of their precision and recall and their running time and conclude the paper by some insights from our comparison. In particular, our experimental results justify a link-based recommender system for weblogs and other social networks with high average degree. Our experimental results can be found in Section 5.

1.2   Related Work

Recommender systems have been extensively studied recently [2, 3, 11]. Recommender systems are closely related to clustering algorithms. Different clustering algorithms for large data sets have been proposed. In this paper, we use ideas from spectral clustering, and clustering based on random walks. A good survey of spectral clustering can be found in a recent paper by Verma and Meila [7]. They have studied several spectral algorithms in their paper and also compared their performance. In particular, they have studied the algorithms by R. Kannan, S. Vempala, A. Vetta [9], and an algorithm by J. Shi, J. Malik [8]. In the context of clustering algorithms based on random walks, the paper of Speilman and Teng [14] introduces a fast algorithm for local partitioning based on refined random walks. Also, a recent paper by Andersen, Chung and Lang [4] gives a local partitioning algorithm using the personalized PageRank.

Recommender systems are also related to ranking algorithms in that we need to rank the recommended items in some order. In this paper, we use ideas from the PageRank Algorithm that has been developed by Page, and Brin [5]. Another important variant of this algorithm is the personalized PageRank which was introduced by Haveliwala [15], and has been used to provide personalized search ranking and context-sensitive search.

2   Preliminaries

In this section, we define the terms that are used throughout the paper.

Weblog Graph: Nodes of the blog graph are the set of weblogs. There is a directed edge from node w to node u if there exists a link from blog w to blog u. As a result, the blog graph is a directed graph. We also study the undirected version of the blog graph in which we make all edges of the directed graph undirected. In other words, we make the graph undirected by putting the backward edges in the graph.

Cut in a graph: A cut $(S, V(G) - S)$ of an undirected graph $G = (V, E)$ is the set of edges from S to $V(G) - S$ [13]. For two sets $A, B \subseteq V(G)$, $\mathrm{Cut}(A, B)$ denote the set of edges from A to B.

Volume of set A : Let $deg(v)$ to be the degree of node v. Then, the volume of set A is denoted by $\text{Vol}(A) = \Sigma_{v \in A} deg(v)$.

Sparsity for two sets: $A, B \subseteq V(G)$ is defined as $\text{Sparsity}(A, B) = \frac{|\text{Cut}(A,B)|}{\text{Vol}(A)\text{Vol}(B)}$ [7].

Connected Component: A connected component of an undirected graph is a maximal set of vertices such that for any two vertices u and v in the set, there is a path from u to v or from v to u.

Strongly Connected Component: A strongly connected component of a directed graph is a maximal set of vertices such that for any two vertices u and v in the set, there is a path from u to v and from v to u.

Random Walk: Given a graph and a starting point, we select a neighbor of it at random, and move to this neighbor, then we select a neighbor of this point at random and move to it etc.. The (random) sequence of points selected this way is a random walk on the graph [12].

Uniform Random Walk: A uniform random walk is a random walk in which at each step we go to each neighbor with the same probability [12].

Eigenvalue and Eigenvector: Let A be a linear transformation represented by a matrix A. If there is a vector $X \in R^n \neq 0$ such that $AX = \lambda X$ for some scalar λ, then λ is called the eigenvalue of A with corresponding (right) eigenvector X.

2.1 Recommender Systems

We assume that each user has a set of favorite items. Favorite items are the items which are already bought, used or chosen. The goal of a recommender system is to recommend a set of items, similar to the set of favorite items.

Recommender systems for weblogs can be designed based on the similarity among the content of the weblogs, the link structure between different weblogs, and the comments posed by the blog owners. Here we briefly explain these three approaches separately.

1. **Content based recommender systems:** Based on the frequent words inside the posts and their titles, we can come up with a set of subject features for each blog. Score of each feature in each blog is determined by the frequency of the words related to this feature. We can view this scoring system of features for weblogs as a ranking system and use this scoring to design a content based recommender system. In order to implement this recommender system we can use one of the known algorithms either model-based or memory-based algorithms. Many such algorithms are described in [1].

2. **Link-based recommender systems:** We can build the blog graph assuming the weblogs as nodes and the links between weblogs as edges of this graph. Clustering this graph can help us reveal similarities among weblogs and as a result, it can be used to design a recommender system. Note that the outgoing links from weblogs can be transferred by users to find the similar and related weblogs, but the main advantage of constructing the blog graph is discovering the similarities between the incoming links for different weblogs. This idea can be used in combination with the content-based

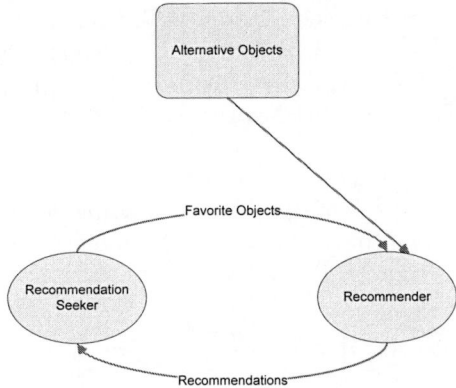

Fig. 1. Recommendation System Model

recommender systems and improve its outcome. We can use the known clustering algorithms such as clustering based on random walks [4, 14], spectral clustering [7], or any type of well-connected components to reveal this information.

3. **Comment-based recommender systems:** Instead of constructing the blog graph, only based on the links between weblogs, we can build the graph based on the posted comments inside the weblogs. The idea is to put an edge from blog A to blog B, if the owner of blog A put a comment (or several comments) for the posts of blog B. We can then cluster this graph which reveals the relation among weblogs.

3   Local Partitioning Algorithms Based on Random Walks

In this section, we present a link-based recommender system based on computing the personalized PageRank of favorite nodes using refined random walks in a social network. In this algorithm, we consider an undirected, unweighted graph $G(V, E)$, where V is the vertex set, E is the edge set, n is the number of vertices and $\deg(v)$ is the degree of vertex v. The algorithm uses a restarting probability distribution vector r. For example, in the PageRank algorithm [5], the distribution is uniform over all nodes. Also, for the personalized PageRank [4, 5], the vector is one at a single vertex and zero elsewhere. Moreover, if we are given a set S of k weblogs for which we want to find relevant weblogs, we can set the restarting vector r as a vector with value $\frac{1}{k}$ on the k vertices corresponding to the set S of weblogs or the user can distribute 1 over k favorite weblogs. Using this restarting probability distribution, we claim that the weblogs that have the largest PageRank values in the following algorithm are more relevant to the weblogs in set S.

Given the restarting vector r, we can run a PageRank algorithm with this restarting probability distribution and report the nodes with high PageRank as the output. The algorithm also uses a restarting probability $p = 0.15$, which is the probability that we restart the random walk from vector r. In fact, we use a variant of random walk which is called the *lazy random walk*. In this variant, at each vertex with probability, ℓ, we remain in the same vertex. Since we are interested in the similarity to our favorite items, using lazy random walks is more appropriate.

We construct the transition matrix A that corresponds to the PageRank random walk with restarting probability p and restarting vector r. $A[i, j]$ is in fact the probability that we will move to vertex j in the graph given that we are in vertex i, that is, for an edge (i, j), we have:

$$A[i, j] = \frac{(1 - p - \ell)}{\deg(i)} + pr(j),$$

for $i = j$,

$$A[i, j] = \ell + pr(j),$$

and for a non-edge pair (i, j):

$$A[i, j] = pr(j),$$

First, we give a rough description of the algorithm. Given the restarting vector r, the restarting probability p, the laziness probability ℓ, and the transition matrix A, the algorithm is as follows:

1. $v = r$.(We start from vector r.)
2. $w = Av$.
3. while $(\text{dist}(w, v) > 0.0001)$ do
   (a) $v = w$.
   (b) $w = Av$.
4. Sort weblogs in the non-increasing order of the value $v(i)$ and output them in this order. If some of the weblogs are already in the list of favorite weblogs or their immediate neighbors, do not output them.

$\text{dist}(w, v)$ in this algorithm is the ℓ_2 distance between w and v.

We would like to improve the running time of the above algorithm to find the b most relevant nodes to a given set S of k favorite nodes in the graph. In particular, we want to design an algorithm whose running time depends on the number b of relevant nodes in the output, and not the size of the whole graph. In other words, we need to design a local way of approximating the personalized PageRank vector of a set of nodes. There are two ways to perform such a task:

Truncating Probability Vectors. The idea of truncating the probability vectors has been developed by Speilman and Teng [14] for local partitioning based on random walks. In this algorithm, after each step of the random walk, we change the entries of the probability distribution vector that are less than a threshold to zero, and then normalize the vector. A natural threshold to use is $\alpha = \frac{\epsilon}{2b}$ where $0 < \epsilon < 1$. [2]. Given the restarting vector $\boldsymbol{r}$, the number of weblogs that we would like to recommend b, a threshold for pruning the resulting vectors α, and a transition matrix A, the algorithm is as follows:

1. $\boldsymbol{v} = \boldsymbol{r}$ (i.e, we start from vector $\boldsymbol{r}$.)
2. $\boldsymbol{w} = A\boldsymbol{v}$.
3. Let q be the number of nodes with positive scores in w.
4. while ((number of nodes with positive scores $\leq b + q$) and ($dist(v, w) > 0.0001$) do
   (a) $\boldsymbol{v} = \boldsymbol{w}$.
   (b) $\boldsymbol{w} = A\boldsymbol{v}$.
   (c) for each $i \in V(G)$ do
      i. if $\boldsymbol{w}[i] \leq \alpha$ then $\boldsymbol{w}[i] = 0$.
   (d) normalize($\boldsymbol{w}$).

In our experiments, we implemented the above algorithm. However, for the sake of completeness, we present an alternative way to perform a locally fast random walk.

Sequentially Pushing Probability from Nodes. The idea of this algorithm has been developed by Andersen, Chung, and Lang [4] for approximating the personalized PageRank vector. The algorithm maintains two vectors $\boldsymbol{p}$ and $\boldsymbol{q}$ that are initialized to $\boldsymbol{0}$ and the restarting vector $\boldsymbol{r}$ respectively. The algorithm then applies a series of push operations moving probability from $\boldsymbol{q}$ to $\boldsymbol{p}$. Each push operation takes the probability from $\boldsymbol{q}$ at a single vertex u, and moves an α fraction of this probability to $p(u)$, and distributes the $1 - \alpha$ fraction of $q(u)$ within $\boldsymbol{q}$ by applying one step of the lazy random walk from u. The algorithm is summarized as the following:

1. Let $\boldsymbol{p} = \boldsymbol{0}$ and $\boldsymbol{q} = \boldsymbol{r}$.
2. While $r(u) \geq \epsilon \deg(u)$ for some node u,
   (a) Pick any vertex u where $r(u) \geq \epsilon \deg(u)$.
   (b) Update p and q:
      i. $p(u) = p(u) + \alpha r(u)$.
      ii. For each outgoing neighbor v of node u, $r'(v) = r(v) + \frac{(1-\alpha)r(u)}{2\deg(u)}$.
      iii. $r(u) = (1 - \alpha)\frac{r(u)}{2}$.

3.1   Random Walk on Directed Graphs

The local partitioning algorithms described in Section 3 are used for undirected graphs and can be applied to the undirected version of the blog graph. The original blog graph is a directed graph. If we apply the ideas from the aforementioned algorithms on directed graphs, one issue is that if the directed graph has a sink, most of the probability measure will go to the sink. In order to apply the ideas from those algorithms on directed graphs, we need to find the strongly connected components(SCC) of the directed graph, and change the restarting

---

[2] In order to avoid infinite loop, we need to have $\alpha < \frac{1}{b}$.

probability based on the place of the nodes in strongly connected components. In this algorithm, we consider a directed, unweighted graph $G(V, E)$, where V is the vertex set, E is the edge set, n is the number of vertices, and outdeg(v) is the degree of vertex v. Using a linear-time algorithm by Tarjan [6], we first find the strongly connected components of this graph. For a node $v \in V(G)$, let $SCC(v)$ be the strongly connected component containing the node v. We can also find connected components of the undirected graph corresponding to the directed graph. This can be found using a linear-time breadth-first-search algorithm. For a node $v \in V(G)$, let $CC(v)$ be the connected component containing the node v. It follows that $SCC(v) \subseteq CC(v)$. For a subset S of nodes, let $SCC(S) = \cup_{v \in S} SCC(v)$ and $CC(S) = \cup_{v \in S} CC(v)$.

Given a set of favorite nodes S, we partition all nodes into five categories:

Category 1. Nodes in the $SCC(S)$.
Category 2. Nodes that have a directed path to $SSC(S)$.
Category 3. Nodes that have a directed path from $SSC(S)$.
Category 4. All the other nodes in $CC(S) - SCC(S)$.
Category 5. Nodes in $V(G) - CC(S)$.

Let Category$^S(i)$ be the set of nodes of category i is given a set of favorite nodes S.

The idea for directed graphs is to change the restarting probability p based on the category of a node for set S. As discussed in Section 3, the set of favorite weblogs S is the support of the restarting vector r. As a result, starting from a set of favorite weblogs S, the probability of getting to any node of category 5 in a random walk of Algorithm PPR is zero. If we perform the random walk on a directed graph, the probability of getting to any node of category 2 and 4 is also zero. Nodes of category 3 are sinks of the random walk, since there is no path from them to $SCC(S)$. To decrease their rank in the, we set the restarting probability of a node $v \in Category^S(3)$ to a larger number, say $p(v) = 0.5$. For all the other nodes $v \in Category^S(1)$, we set the restarting probability to $p(v) = 0.15$. We also put more probability mass on edges to the same strongly connected component. Let $w : V(G) \rightarrow V(G)$ be a weight function defined as follows: $w(u, v) = 0$ if $(u, v) \notin E(G)$. Otherwise, we let $w(u, v) = 1$ if $v \in SCC(u)$ and $w(u, v) = 0.3$ if $v \notin SCC(v)$.

As a result, we construct the transition matrix A that corresponds to our non-uniform random walk with restarting probability p and restarting vector r. $A[i, j]$ is the probability that we will move to vertex j in the graph given that we are in vertex i, that is, for an edge (i, j), we have:

$$A[i, j] = \frac{(1 - p(i) - \ell)w(i, j)}{\sum_{t:(i,t) \in E(G)} w(i, t)} + p(i)r(j),$$

for $i = j$,

$$A[i, j] = \ell + p(i)r(j),$$

and for a non-edge pair (i, j):

$$A[i, j] = p(i)r(j).$$

Given the above transition matrix, we can run the truncated random walk algorithm discussed in Section 3 on the directed graph to compute the approximate personalized PageRank of the nodes.

4   A Spectral Method

We use the intuition behind the spectral clustering to compute a *similarity metric* for the purpose of our recommender system. Recall that the sparsity for two sets $A, B \subseteq V(G)$ is defined as $\text{Sparsity}(A, B) = \frac{|\text{Cut}(A,B)|}{\text{Vol}(A)\text{Vol}(B)}$. A goal of clustering is to find cuts for which the sparsity is small. Spectral clustering is suitable for clustering graphs in which the maximum sparsity is not large. Spectral clustering is considered to be a hierarchical and global clustering. Hence, in order to use spectral clustering for our system, we need to change the current known algorithms for spectral clustering. The idea is to use the eigenvectors of a normalized adjacency matrix to evaluate the distance between nodes of a graph. In order to recommend a set of weblogs relevant to a favorite blog w, we can output the weblogs that have small distance to blog w according to a similarity metric computed based on the eigenvectors of the largest eigenvalues. Intuitively, each eigenvector of a large eigenvalue represents one type of clustering the data into two clusters. The importance of an eigenvector is directly related to its corresponding eigenvalue, i.e., a larger eigenvalue indicates a better partitioning of the graph.

There are several variants of the spectral clustering algorithms in the literature. A good survey on this topic is by Verma, and Meila [7]. They categorize spectral algorithms into two major sets: (i) *Recursive Spectral*, in which, we partition data into two sets according to a single eigenvector and by recursion we generate the specified number of clusters, or (ii) *Multiway Spectral*, in which, we use the information hidden in multiple eigenvectors in order to directly partition data into the specified number of clusters. In particular, our algorithm is based on the ideas for the recursive spectral algorithm proposed by Semi and Malik [8], Kannan, Vempala, and Vetta [9], and the multiway spectral algorithm of Ng, Jordan, and Weiss [10].

Let A be the adjacency matrix of the blog graph, and D be a diagonal matrix such that $D[i, i] = \deg(i)$ for a node i. Given a favorite blog w for which we want to find similar weblogs, our algorithm is as follows:

1. Calculate $P = D^{-1}A$
2. Let $1 = \lambda_1 \geq \lambda_2 \geq \ldots \geq \lambda_n$ be the eigenvalues of P, compute $v^2, v^3, \ldots, v^t$ which are the eigenvectors corresponding to $\lambda_2, \lambda_3, \ldots, \lambda_t$.
3. Find a set of weights $c_2, \ldots, c_t$ for each eigenvector. c_i shows how valuable or important vector v^i is for clustering purposes. Intuitively, if $\lambda_i \leq 0$, $c_i = 0$, and for $\lambda_i > 0$, $c_i = f(\lambda_i)$ where $f(x)$ is an increasing convex function, for example $f(x) = x^2$, or $f(x) = e^x$.
4. For any blog j and any vector v^i, let the difference $d_i(j) = |v^i(j) - v^i(w)|$.
5. Sort the vertices in the non-decreasing order of the value $q_j = \sum_{2 \leq i \leq t} c_i d_i(j)$ and output the vertices in this order.

The vector q is the distance measure to node w. Considering this distance measure for all nodes results in a similarity metric over all pairs of nodes.

Since computing the eigenvalue and eigenvectors require matrix multiplication, the running time of the above algorithm is at least as bad as the matrix multiplication algorithm[3].

However, we could approximate the eigenvalues and eigenvectors of the graph and use them to define the metric. In our experiments, we improve the running time of this algorithm by computing a *candidate set* of 1000 weblogs related to our favorite blog, and then use the spectral algorithm to rank weblogs in the candidate set. In order to find this candidate set, we could use the content information of weblogs or use some local partitioning algorithm that finds all weblogs that are relevant to a particular blog. In our experiments, we compute the truncated random walk algorithm that computes the approximate personalized PageRank of each blog with respect to the favorite blog. As a candidate set of weblogs, we output 1000 weblogs with highest approximate personalized PageRank with respect to the favorite blog. Then we can find the above metric based on the eigenvalue and eigenvectors of the induced graph of the candidate set to rank the nodes of this set.

5   Evaluation

Evaluating recommender systems and their algorithms is inherently difficult for several reasons. First, different algorithms may be better or worse on different data sets. Many collaborative filtering algorithms have been designed specifically for data sets where there are many more users than items. Such algorithms may be entirely inappropriate in a domain where there are many more items than users. Similar differences exist for ratings density, ratings scale, and other properties of data sets. The second reason that evaluation is difficult is that the goals for which an evaluation is performed may differ. Finally, there is a significant challenge in deciding what combination of measures to use in comparative evaluation [11].

5.1   Implementation

In order to evaluate the performance of the algorithms, we implemented a crawler and made a dataset. The crawler is a focused crawler meaning that it only crawls weblogs which are not in the weblogs' domain. We constructed the blog graph for the dataset and performed four algorithms on it:

1. Approximate personalized PageRank on directed graphs (DPPR).
2. Approximate personalized PageRank on undirected graphs (PPR).
3. Spectral algorithm applied on the a candidate set from DPPR (SDPPR).
4. Spectral algorithm applied on the a candidate set from PPR (SPPR).

---

[3] The best known algorithm for matrix multiplication is $O(n^\alpha)$ where $\alpha = 2.37$.

As discussed in Section 3, we have two options for implementing the local computation of approximate personalized PageRank: the truncated random walk, and pushing probability from nodes. We used the truncated random walk algorithm for our implementation and achieved improved running time. Since the spectral algorithm is not based on a local computation, we run the PPR and DPPR algorithms to find a set of 2000 candidate nodes, and then compute the spectral metric for this set of nodes, and rank them based on the distance of these nodes in this metric. In order to find the set of candidate nodes to feed to this spectral algorithm, one can use other types of content-based algorithms as well.

5.2 Metrics

For the system we are going to develop, we adapt two metrics defined in information retrieval, *precision* and *recall*. In this case we define precision and recall respectively as follows:

$$\text{Precision} = \frac{|\text{Recommended weblogs} \cap \text{Favorite weblogs}|}{|\text{Recommended weblogs}|},$$

$$\text{Recall} = \frac{|\text{Recommended weblogs} \cap \text{Favorite weblogs}|}{|\text{Favorite weblogs}|}.$$

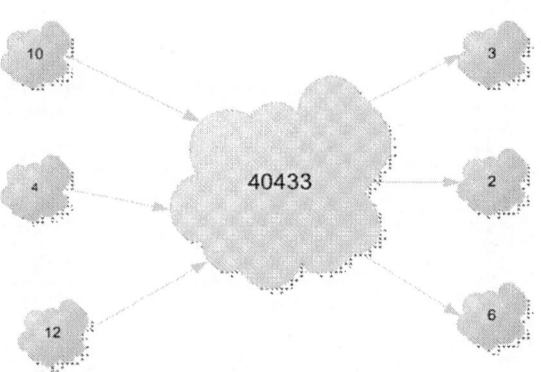

Fig. 2. An abstract model of our blog graph: strongly connected components

5.3 Experimental Results

In order to perform such evaluation, we omit a portion of links from the list of outgoing links of a blog and reconstruct the graph without that information. Then, we apply the algorithm on the new graph and see how many omitted links are rediscovered, then we calculate the metrics.

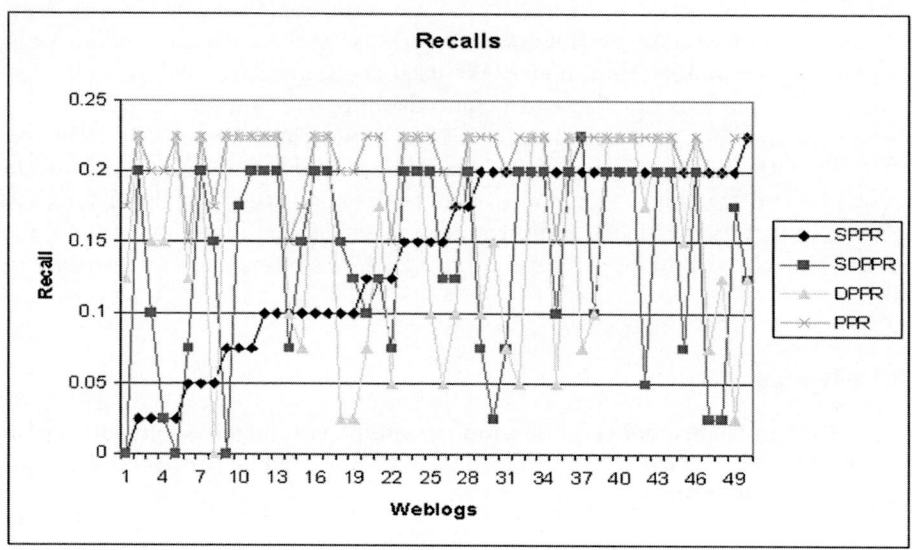

Fig. 3. Recalls

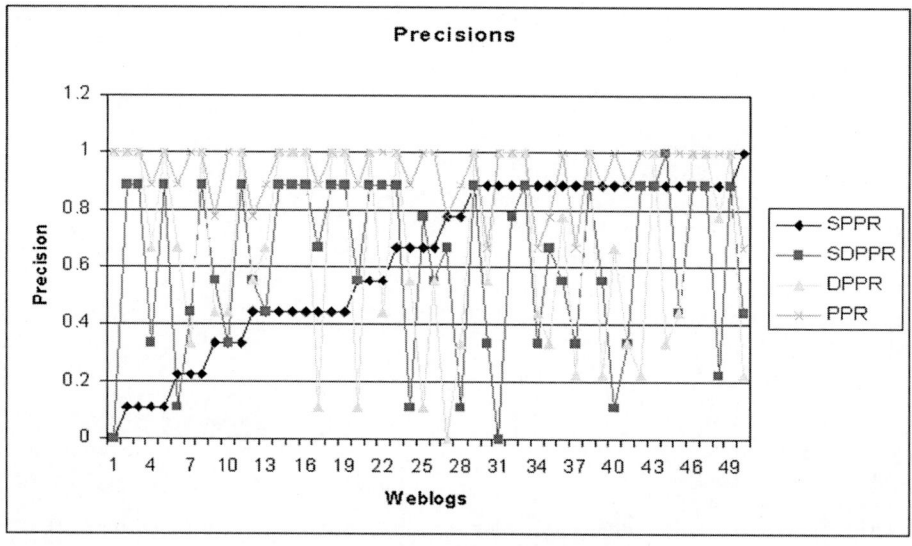

Fig. 4. Precisions

Our dataset contains 120000 weblogs, which have links to about 4000000 other weblogs. The average degree of the blog graph is 50. This average degree is more than the typical average degree of nodes of other social networks like the web graph. This implies that there is more information hidden in the link structure of

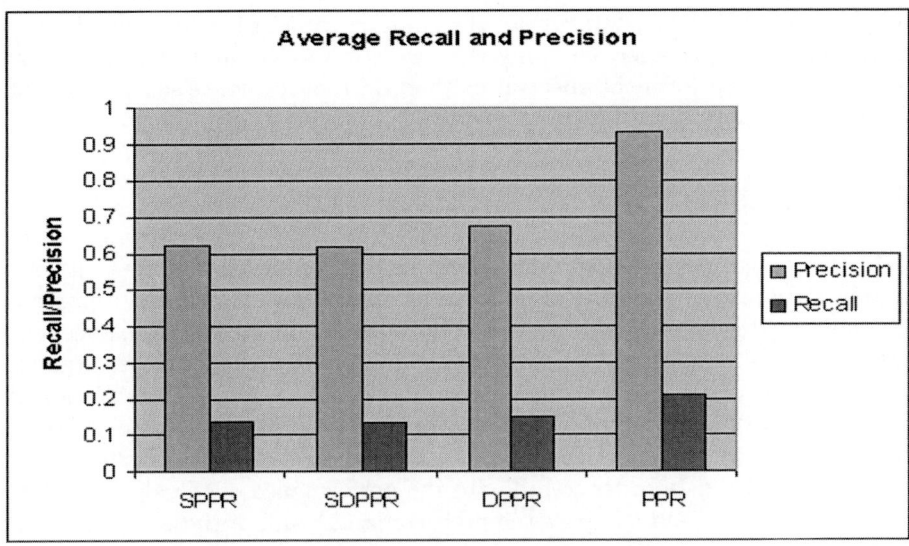

Fig. 5. Comparing average of recalls and precisions

this graph compared to other social networks. As we will see, this might be the main reason that our algorithm performs very well on this graph. Moreover, as we expected the blog graph has a large strongly connected component containing 40433 nodes and many small strongly connected components.

In order to have meaningful experiments, we select 50 high-degree weblogs at random. The degree of each of these weblogs were at least 50, and we removed 10 to 15 outgoing neighbors of each of these 50 weblogs. The recalls and precisions calculated for these 50 randomly chosen weblogs are plotted in Figures 3 and 4 respectively. The average of recall and precision for all four algorithms are also plotted in Figure 5. As it can be seen, all the four algorithms have acceptable performance. The performance of these four algorithms is ordered as follows: personalized PageRank algorithm on undirected graphs, personalized PageRank algorithm on directed graphs, Spectral algorithm applied on the a candidate set from DPPR, Spectral algorithm applied on the a candidate set from DPPR. These experiments justify the use of link structure for a recommender system for weblogs, and perhaps other social networks with high average degree. Moreover, it proves the applicability of local partitioning algorithms to estimate the personalized PageRank algorithm.

Running Time. Both PPR and DPPR algorithms are based on local computation of approximate personalized PageRank vectors and run very fast. We implement these two algorithms in C. In comparison, the running time of the algorithm on directed graphs is faster. The reason is that the outdegree of vertices is smaller than its degree in the undirected graph and as a result we deal with larger vectors in PPR.

The time to construct the similarity metric based on the spectral method for the whole graph is much worse than the running time of the local partitioning algorithms, however, since we apply the algorithm on a candidate set of 1000 nodes and then perform the spectral method on the candidate set, the running time is comparable. The spectral algorithm is implemented in Matlab.

6   Conclusions

In this paper, we develop two main ideas for a recommender system based on the link structure of the blog graph. We examine these two ideas on directed and undirected graphs. One idea is to use the personalized PageRank vector to capture the relevance of nodes on a graph and apply the local truncated random walk algorithms to approximate this vector. We also extend the random walk to a refined random walk on directed graphs. The other idea is to introduce a metric based on the spectral methods that can be used to rank the relevant nodes in a network. In order to evaluate the performance of the algorithms, we construct a dataset and compare the precision and recall of these methods on a sample subset of this dataset.

Our experimental results show that the performance of all algorithms are acceptable for weblogs. The main reason for this good performance is the high average degree of the blog graph which indicates that one can achieve a reasonable performance for a recommender system for weblogs by only using the link information of the blog graph. This observation implies that for social networks with high average degree using the link structure for a recommender system is a realistic approach.

The local approximate personalized PageRank algorithm has a better performance compared to the spectral method for our data set. The running time of the personalized PageRank algorithm is also much better than the spectral method that we used. The results of the spectral method are mainly used as a base for comparison, and justify the use of the random walk algorithm. Between the two personalized PageRank algorithms on directed and undirected graphs, the algorithm for directed graphs run faster, but the performance of the algorithm for undirected graphs is better.

References

1. http://www.blogpulse.com (visited, December 2006)
2. Parsons, J., Ralph, P., Gallagher, K.: Using viewing time to infer user preference in recommender systems. In: AAAI Workshop in Semantic Web Personalization, San Jose, California (July 2004)
3. Resnick, P., Varian, H.: Recommender Systems. Communications of the ACM 40, 56–58 (1997)
4. Andersen, R., Chung, F., Lang, K.: Local Graph Partitioning using PageRank Vectors. In: Proceedings of the 47th Annual IEEE Symposium on Foundations of Computer Science (FOCS 2006), pp. 475–486 (2006)

 5. Brin, S., Page, L., Motwani, R., Winograd, T.: The PageRank citation ranking: Bringing order to the web, Technical report, Stanford Digital Library Technologies Project (1998)
 6. Tarjan, R.E.: Depth-first search and linear graph algorithms. SIAM Journal on Computing 1(2), 146–160 (1972)
 7. Verma, D., Meila, M.: A comparison of spectral clustering algorithms. Technical report UW-cse-03-05-01, University of Washington
 8. Shi, J., Malik, J.: Normalized cuts and image segmentation. IEEE Transactions on Pattern Analysis and Machine Intelligence 22(8), 888–905 (2000)
 9. Kannan, R., Vempala, S., Vetta, A.: On clusterings- good, bad and spectral. In: Proceedings of the IEEE Symposium on Foundations of Computer Science (FOCS 2000), pp. 367–377 (2000)
10. Ng, A., Jordan, I., Weiss, Y.: On Spectral Clustering: Analysis and an algorithm. Advances in Neural Information Processing Systems 14, 849–856
11. Herlocker, J.L., Konstan, J.A., Terveen, L.G., Riedl, J.T.: Evaluating Collaborative Filtering Recommender Systems, ACM Transactions on Information Systems (TOIS) (2004)
12. Lovasz, L.: Random walks on graphs: A survey (January 1993)
13. Cormen, T.H., Leiserson, C.E., Rivest, R.L., Stein, C.: Introduction to Algorithms, 2nd edn. (2001)
14. Spielman, D.A., Teng, S.: Nearly-linear time algorithms for graph partitioning, graph sparsification, and solving linear systems. In: ACM STOC 2004, pp. 81–90. ACM Press, New York (2004)
15. Haveliwala, T.H.: Topic-sensitive PageRank: A context-sensitive ranking algorithm for web search. IEEE Trans. Knowl. Data Eng. 15(4), 784–796 (2003)

Author Index